THE
OLYMPIC DAM
STORY

HOW WESTERN MINING DEFIED THE ODDS TO DISCOVER AND DEVELOP THE WORLD'S LARGEST MINERAL DEPOSIT

DAVID UPTON

Copyright © David Upton 2010

The moral right of the author has been asserted.

All rights reserved. No part of this book may be reproduced, stored in or introduced into a retrieval system, or transmitted in any form or by any means (electronic, mechanical, photocopying, recording or otherwise) without the prior written permission of the publisher.

National Library of Australia Cataloguing-in-Publication entry:

Upton, David (David Gregory), 1964 —

The Olympic Dam Story: How Western Mining defied the odds to discover and develop the world's largest mineral deposit

978-0-646-54381-9 (pbk)

Includes index

Western Mining Corporation
Mines and mineral resources — South Australia — History
Mineral industries — South Australia — History
Olympic Dam Mine (S. Aust)

338.20099423

First published in 2010 by UPTON Financial PR,
P.O.Box 8430, Armadale, Victoria, Australia 3143
Email: davidupton3@bigpond.com

Contents

Acknowledgments · v
Foreword · vi
Chapters
 1 A quick tour of Olympic Dam · 1
 2 How big? · 11
 3 How did it get there? · 17
 4 Driven to explore · 25
 5 Modern leaders · 37
 6 Reinventing the search for copper · 53
 7 Homing in on Andamooka · 65
 8 Olympic Dam becomes the target · 87
 9 Discovery — at last · 101
 10 A huge deposit revealed · 121
 11 Accidental elephant? · 127
 12 The quietly achieving partner · 133
 13 Drilling out · 143
 14 Politics and protests · 155
 15 Epilogue · 165
Index · 173

Acknowledgments

I was fortunate that so many of the individuals involved in the discovery and development of Olympic Dam were enthusiastic about sharing their personal records and recollections. They were always ready to respond to my many questions and requests for assistance.

My sincere thanks to (in alphabetical order) Tom Allison, Gavan Collery, Ken Cross, John Emerson, Dan Evans, Douglas Haynes, Keith Johns, Jim Lalor, Hugh Morgan, Don Morley, Henry Muller, Sir Arvi Parbo, Gilbert Ralph, John Reynolds, Hugh Rutter, Kym Saville, Richard Schodde and Roy Woodall.

Special thanks goes to Sir Arvi, who inspired the book and has been tireless in helping me and others record the history of mineral exploration in Australia.

Thank you to BHP Billiton for providing access to its archives, which contain the official records carefully compiled by Western Mining over many decades, and to the Department of Primary Industries and Resources South Australia.

Lastly, a heartfelt thank you to my wife, Stacey, for her advice and support, which were invaluable, as always.

Foreword

Minerals and energy have made an indispensable contribution to the betterment of human life since the beginning of civilisation.

The demand for minerals and energy is increasing as more than half of today's world population work their way out of poverty. Additionally, this population is predicted to increase by about one third before reaching a plateau later this century. The demand will be affected by economic cycles, as it has been in the past, but the trend will be upwards.

This demand can be partially met by improvements in technology and/or higher prices making lower grade deposits economic and by recycling, but it will be essential to discover major new deposits. In poorly explored parts of the world such finds may still be made at or near the surface, but this is increasingly unlikely in Australia. In this country the main potential for future discoveries is at greater depth.

It is in the nature of minerals exploration that successes are rare. Very few prospects become orebodies. Geosciences and exploration technology are continually improving, but deeper discoveries will be even more difficult and more expensive to make than the shallower finds in the past.

The Olympic Dam copper-uranium-gold-silver deposit in South Australia is a very large orebody, with no surface signs of its existence and covered by 350 metres of barren rock. Thirty five years after discovery, the limits of the mineralisation have not yet been determined.

David Upton tells the Olympic Dam story vividly and in an eminently readable manner. A geologist by training, he has the rare ability of explaining the scientific and technical issues simply. He has interviewed many of the people who were there and includes their recollections. We are greatly indebted to David for this thoroughly researched and detailed record of the discovery and its development.

Readers will draw their own conclusions, but to me several aspects stand out. The first is that writers, poets, composers and painters can create masterpieces on their own and Albert Einstein can take sole credit for having thought of the Theory of Relativity, but no-one in this industry can achieve anything without teamwork by many others. Leaders are important in setting the goals, assembling the means and encouraging those in the field, but ultimately success depends on the many capable, well-trained and dedicated people working together as a team who are not discouraged by the inevitable setbacks.

Secondly, in the minerals industry it is necessary to think and act very long term. Perseverance and the ability to continue through the periodical downturns are essential. Western Mining searched for copper for 20 years before a drillhole intersected copper mineralisation at Olympic Dam. It took another 13 years to produce the first product. If the reported plans of the present owner, BHP Billiton, are implemented, full production may not be reached until nearly 50 years after discovery. I believe that Olympic Dam will still be producing more than a hundred years from now.

Finally, there have been numerous minerals finds in Australia in the last 35 years, but nothing comparable to Olympic Dam. The discoveries in the 1960s and 1970s were massive, have been extended by further exploration and have made the industry a world leader, but large new discoveries must be made for this to continue.

High level and persistent exploration effort is necessary if the Australian minerals industry is to maintain its world leadership position. I hope *The Olympic Dam Story* will assist in understanding that there is nothing easy or granted about it.

It is absolutely vital that minerals exploration is welcomed, encouraged, and made worthwhile. Exploration is high risk, unglamorous, and a hard slog, but the future depends on it. The rewards from the successes must compensate for all the unsuccessful exploration, for the long lead times from discovery to cash flows from production, and for the poor returns during the periodical downturns.

A strong and successful minerals industry has been essential to Australia in the past and continues to be in the interests of every Australian, now and in the future.

Sir Arvi Parbo
Melbourne, October 2010

A QUICK TOUR OF OLYMPIC DAM

Aircraft were an unusual sight on the strip at Roxby Downs pastoral station, more than 500 kilometres north of the South Australian capital of Adelaide. In fact, the only real flight activity all year was during livestock musters around March and September, when station owner Tom Allison would be up and down several times a day in his single-engined Cessna.

Pastoralists further south scoffed when he built the airstrip and bought a plane after taking over the property in 1968, but Allison discovered that herding livestock from the air was the only way to work Roxby Downs. The scrappy vegetation on a square kilometre of land might feed six sheep a year. Allison had 2,000 square kilometres to cover and up to 15,000 sheep. In addition to the vast spaces, Allison had to contend with a maze of red sand dunes that stretched west to east across large areas of Roxby Downs. "It would take me a week on a motorbike to round up 300 sheep in the sand dunes. I could do the same job in three hours in a light plane," Allison says.

Outside of mustering season, many weeks could pass without an aircraft engine drowning out the distant sounds of sheep and native pines in the wind. Today would be different. On 5 November 1976, a twin-engined Baron droned in from the south after lifting off from Adelaide Airport about 90 minutes earlier. Aboard were some of Western Mining's most senior executives, including the director of exploration, Roy Woodall, the head of mineral exploration in eastern Australia, Jim Lalor, and the most senior man in the company, Arvi Parbo (later Sir Arvi), the chairman and managing director.

The presence of top-ranking company men might normally suggest a great discovery had been made, but there was no such news bringing this group to Roxby Downs. Parbo, the Estonian migrant who had already become one of Australia's top businessmen, was always eager to get out of Melbourne head office and see what the exploration geologists were doing.

A QUICK TOUR

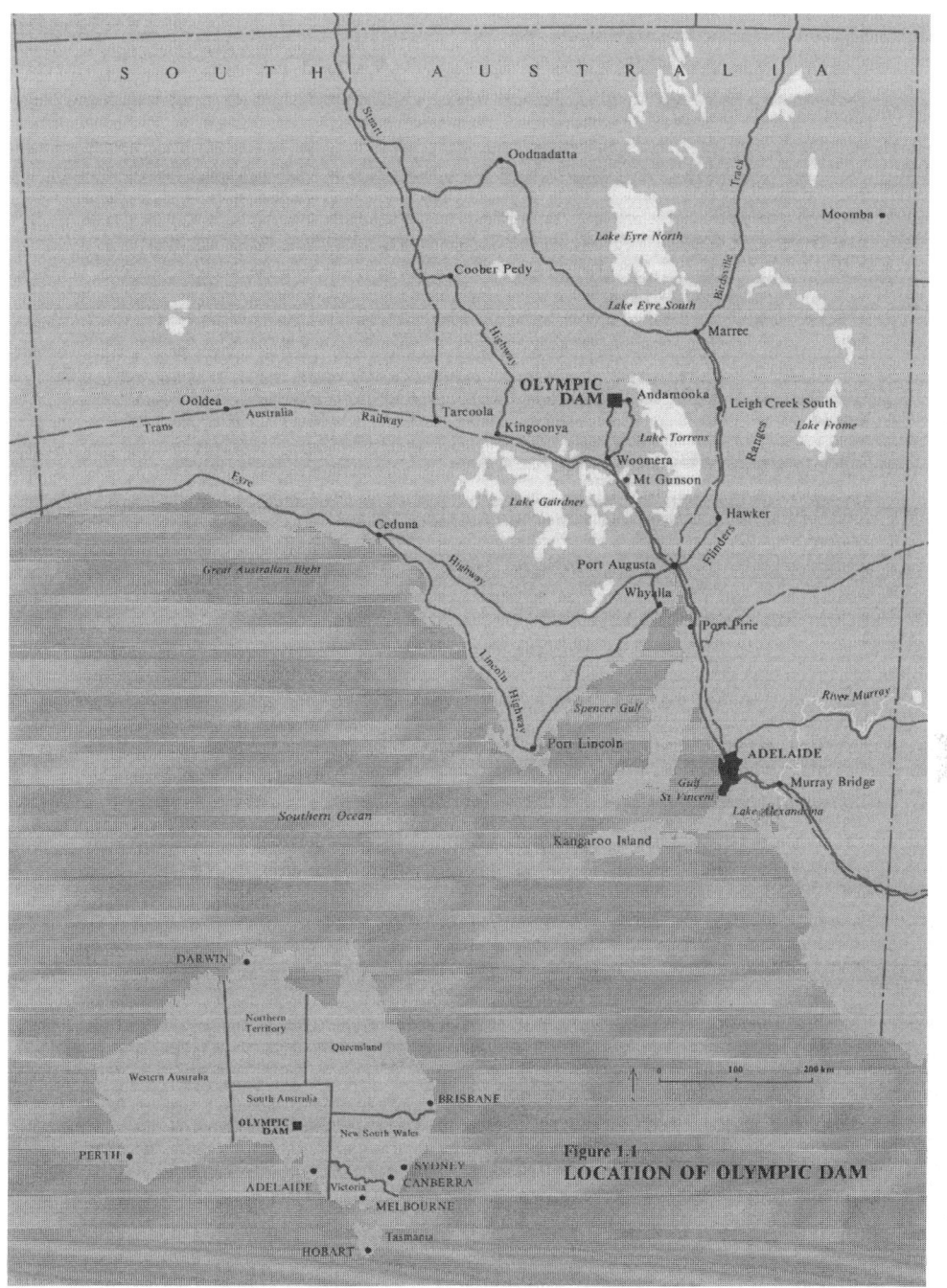

Location map from 1982 published by Western Mining and its new partner, BP

Woodall had called a few weeks earlier and invited him on a trip to the Pedirka Basin in far north South Australia, where a new oil exploration program had begun. He suggested a stop-over at Roxby Downs station to look at the latest drill core from the company's copper exploration project.

The company was using a costly drilling rig with a diamond-studded bit to cut a 36.5 millimetre core of rock to depths of around half a kilometre. Drilling was slow, perhaps 10 to 12 metres a day. Every day, the driller lifted new core from the deepening hole, broke it into short lengths and placed the pieces in steel trays of U-shaped grooves. He would carefully log the below-ground depth of every section for later inspection by the company's geologists back in Adelaide. Big mining companies had geologists sitting around on rigs for weeks at a time, but Western Mining could not afford such luxury.

The project was now drilling its tenth exploration hole (Roxby Diamond 10 or RD10) and had still not made a convincing find. The excitement of more than a year earlier, when the first hole discovered sub-economic grades of copper, had long faded. Falling copper prices made the copper exploration program an even greater test of nerves. However, Parbo had faith in the company's innovative geoscientists. Their new ideas and expertise had led to major discoveries in each of the past three decades, and he would keep supporting the copper project as long as they recommended.

He did not know that even some of the geologists were starting to doubt the science that had led them to the middle of a region where not even one deep exploration hole had been drilled in an area as big as Tasmania. They hoped to find deeply buried basalt rocks altered by volcanic fluids to a burnt, rusty red from the distinctive blue colour seen in the stone buildings and street paving of Melbourne.

A recent PhD by one of their geologists, Douglas Haynes, predicted that, if they could find altered basalts, a copper deposit might be trapped nearby in overlying sedimentary rocks. Instead, the first exploration hole, RD1, found a rock none had seen in all their careers or studies at top universities around the world. At first, some of them called it basalt because it was dark and heavy, but they soon conceded their hearts had overtaken their minds.

This unknown rock was mostly the jagged fragments of a granite that had been blown to bits. The explosive force was so powerful that some of the pieces or clasts were as big as a family home. The rock had later soaked in iron-rich, volcanic fluids at temperatures of up to 400 degrees Celsius. The rocks were a mystery, but at least this unusual exploration venture had found some copper. Perhaps it might yet find higher concentrations of copper minerals, and in large enough tonnages to justify a mine.

Lalor clearly recalls the flying visit to Roxby Downs Station. "The company didn't allow landings on rough dirt strips if Arvi was on board. We normally landed on Lake

Blanche, a dry salt lake that wasn't even recognised as a certified landing strip. The nearest strip we could use on this visit was the one at Roxby Downs Station. This was 30 kilometres from RD10, and we just wouldn't have time to take Arvi to the drill site and back again. So I asked the drilling foreman, John Emerson, to have a couple of trays of core sample from RD10 available to inspect at the airstrip.

"We landed, and John was there. I looked down the airstrip and there was row after row of core tray laid out. They had carted it all in. I said 'John, what are you doing? I told you just to put out two or three different trays!' He said 'Just have a look.' Well, it was unbelievable. There was 200 metres of core laid out and you could see it was going to go two or three percent (copper). It was just so exciting. I had never seen drill core like that, and most people probably never would," Lalor says.

An exploration geologist could make a career out of a copper intersection a fraction of the length of this one; 200 metres with shiny copper sulphides was the stuff of dreams.

One of the greatest mineral discoveries ever had sat unrecognised for weeks under South Australia's far north sky, growing more magnificent every day by 10 or so metres. RD10's discovery was unappreciated until Emerson arrived at the drill site the previous day with his boss, Eric Steart, the Kalgoorlie-based drilling superintendent who by chance was also in the area. Emerson was no geologist, but as an experienced drilling hand from the company's nickel and gold mines in Western Australia, he knew the significance of the sulphides now visible over hundreds of metres.

Sir Arvi recalls a lot of good-natured banter about the discovery being organised for that day's visit. Within an hour or so, the group flew on to Oodnadatta to set out by road on their tour of the company's oil exploration activities. That evening, back at the Continental Hotel, celebrations and games of snooker went late into the evening with anyone who cared to join in. The next morning, the hotel's cook was in no state to begin breakfast, taking precious time out of the day's schedule for the town's important visitors.

With the touchdown at Roxby Downs airstrip the previous day, Olympic Dam changed from a bold exploration experiment to a probable mine, and a huge one at that. It would take Western Mining almost another two years to be confident it actually had an orebody. Later drilling would show RD10 had intersected an "outlier" — a small pocket of copper mineralisation that was more than a kilometre from the main ore body. But there was never any question of giving up after the massive intersection at RD10.

Development challenges

Western Mining eventually brought the deposit into production, but it would take another 13 years. If the company had found only copper, it might have taken five years.

But it had also found uranium interlocked with copper, gold, silver and other minerals within a very complex ore body.

The Labor Premier of South Australia, Don Dunstan, was opposed to uranium mining. The Federal Government of Gough Whitlam had frozen uranium development while an enquiry was held into its safety, which had continued under the new Government of Malcolm Fraser. Even if Olympic Dam could skirt these issues by being deemed a copper/gold mine, it was prohibited from selling the uranium that would inevitably be produced by the extraction of copper and gold.

Government leaders of the day hoped a metallurgical miracle would save them from the thorniest of questions. Do we really have to stop the mine because of the uranium content? Surely the copper and gold could be mined and the uranium left in the ground.

But the economic minerals were intimately mixed, and Western Mining would need to go to extraordinary lengths to produce saleable copper and gold, free of any taint of radioactivity. It designed and built a metallurgical and processing plant unique in the world. It even had to build its own copper refinery in the middle of the Australian outback — an unthinkable idea in any other circumstances. If the company was forced to stockpile the separated uranium oxide after all the costs of removing it from the copper and gold, Olympic Dam would never be able to compete with conventional copper mines around the world. There was no practical or economic way to selectively mine Olympic Dam.

Anti-uranium sentiment was running high in the 1970s and would reach a crescendo in 1979 with the release of *The China Syndrome* and an accident at the Three Mile Island nuclear power plant in Harrisburg, Pennsylvania. But pro-development forces were also growing strong. South Australia was in severe economic difficulty. The manufacturing industries on which it built prosperity in the 1950s and 1960s were in recession and the state had the highest unemployment rate on the Australian mainland. Projects such as Olympic Dam and the Cooper Basin Liquids Project were regarded as essential.

Political and community support for developing Olympic Dam was eventually won in 1982, but by the slimmest of margins. State legislation permitting the mine to go ahead (the Roxby Downs Indenture Bill) was passed only after a former official of the Waterside Workers Union and Labor Party stalwart, Norm Foster, crossed the floor in South Australia's Legislative Council on 18 June 1982 to vote with David Tonkin's Liberal Government. The bill was defeated in a similar vote just a day earlier, but this time Foster resigned from his party and rebelled against its position on Olympic Dam.

Tonkin set the scene for these dramatic events by declaring he would call an election if the bill was defeated a second time. He was swept into office in 1979, shortly after the end of the decade of Labor Premier Don Dunstan, with a mandate to get going with Olympic Dam. In the days leading up to the vote, Tonkin took the unusual step of appearing in a series of television advertisements to push the pro-development argument.

Passage of the bill gave Western Mining and its new partner, BP, the certainty they needed to invest almost $2 billion in today's dollars ($800 million in 1985) to develop the mine and the complex processing plant. They would also invest heavily with the state government in building the new town of Roxby Downs to house the mine's workforce.

Olympic Dam not only challenged governments, it also tested Western Mining. The company had enjoyed legendary exploration success, but this time it found something much bigger than the company's balance sheet could handle.

At an early stage, the South Australian Government was urged to nationalise Olympic Dam on the advice of the influential former chief of the Department of Mines, Sir Ben Dickinson. While this seems like an extreme position today, Australian governments on both sides of politics in the 1970s were not shy about getting involved in uranium mining and marketing. The Federal Minister for Minerals and Energy, Rex Connor, had brought down the Whitlam Government in 1975 because of bungled attempts to finance massive government investment in energy, not because of the plans *per se*.

The looming threat from the South Australian government accelerated Western Mining's plans to find a development partner. Potential candidates were plentiful because of the global scale of the discovery and the plans of many of the world's giant oil companies to diversify into minerals at that time.

BP would become Western Mining's partner in 1979, earning a 49% interest in Olympic Dam in return for financing the project. BP departed almost 20 years ago and is largely forgotten, but its contribution was crucial. It committed to spend more than $300 million in today's terms on the feasibility of a mine that at the time was not even allowed by the State Government. The gamble paid off, but the big rewards were a long way off.

The mine was finally commissioned in June 1988 at a size not much larger than a pilot plant, producing 45,000 tonnes of copper, 1,000 tonnes of uranium oxide and about 70,000 ounces of gold per annum.

Western Mining and BP did not want to risk more than the minimum capital required to get Olympic Dam into production. Government approval was narrowly won and the partners were still nervous about the political risks.

Large technical risks were also still ahead. The Olympic Dam orebody is incredibly complex and unlike anything else in the world. It would be a major achievement just to build Olympic Dam and establish the mine as a reliable supplier. This was a particularly important factor for the notoriously conservative nuclear power utilities that would become buyers of Olympic Dam's uranium.

Against this background, the partners decided to develop Olympic Dam as an underground mine. While it was technically possible to mine the ore body from the surface by open-cut methods, this would require a much greater initial investment. It would also mean years of digging through 350 metres of barren sediments before ore could be extracted and one dollar of cash generated from sales of processed minerals.

The mine was expanded in a series of steps; to 60,000 tonnes of copper per annum in 1992 and 84,000 tonnes in 1993. In the same year, BP sold its interest back to Western Mining as part of its exit from the minerals business worldwide.

In 1996, work began to increase production to 200,000 tonnes of copper per annum. This was a major expansion that required almost as much investment as Western Mining had already sunk into the development. The expansion was completed in 1999.

Six years later, BHP Billiton succeeded in a $9 billion takeover bid for Western Mining, with Olympic Dam regarded as the clear prize. A few months earlier, the company had become the target of a hostile takeover bid from Xstrata, an aggressive global resource company, based in Switzerland. Xstrata had recently taken over MIM, the venerable Australian mining company of Mt Isa, Queensland, in another fiercely contested corporate battle.

When BHP Billiton announced a rival takeover bid for Western Mining, there was relief the company would be saved from foreign ownership, but it was still the end of an era that began with a small gold mining operation in 1933. The company's exploration achievements over many decades, most notably Olympic Dam, had been a source of myth and wonder in Australia and around the world.

BHP Billiton wasted no time in taking a fresh look at Olympic Dam. The change in ownership coincided with a surge in prices for copper and uranium, which slid backwards during Western Mining's ownership over the previous two decades.

BHP Billiton's non-ferrous metals division, under future managing director, Marius Kloppers, set about finding how much could be done with Olympic Dam in a China-led resources boom and by a company with a much bigger balance sheet than Western Mining. It was as though he had found a copy of Sir Arvi's speech at the 1988 opening of the mine, in which he predicted it would be a resource with a life of several hundred years.

For the next two years, BHP Billiton had 20 diamond drilling rigs working around the clock to unlock the secrets still held by an orebody that continued to defy all the rules of economic geology. It concluded that Olympic Dam, even after 20 years of operation, was the second most valuable ore body in the world, behind only the Norilsk nickel and copper mines of Siberia. It has since announced two further increases in the size of the resource to 9,000 million tonnes, almost five times the size of Western Mining's first estimate in 1982.

BHP Billiton has also announced plans for a six-fold increase in the amount of ore extracted every year through the creation of the world's largest open-cut mining operation. The plans envisage it would take from 2011 to 2017 just to dig away the overlying waste rocks and access the deeply buried ore.

If the expansion is approved and runs to schedule, open-pit extraction of ore will begin more than 40 years after that fateful touch down on the quiet grass strip at Roxby Downs station in 1976. Mining is projected to continue for at least another 40 years. Those who know the mine say that in all likelihood it will have a life that runs into many hundreds of years. Even BHP Billiton has still not found the limits of the orebody.

Back to the beginning

There are many reasons to marvel at Olympic Dam, but none surpass the story of its discovery by an eclectic group of geoscientists at Western Mining, most of whom were in their 20s or 30s. How did they find Olympic Dam where no-one had previously dared to imagine? It did not even outcrop at the surface, unlike almost every other major mineral discovery in history.

Decades later, it is hard to grasp how radical they were, but a great deal is revealed by the actions of Western Mining's competitors in the years following the discovery.

Competitors from around the world rushed to explore any available ground near Western Mining's mineral licences after the spectacular results from RD10 were announced on 18 November 1976. These newcomers spent millions searching for the wrong target because they knew so little about the discovery or the nature of the mineralisation. They did not contemplate that Western Mining's "pay dirt" was the hard rocks in the geological basement at depths below 350 metres. Everything they had learned about copper deposits told them Western Mining must have hit its fabulous copper intersections in the shallower sedimentary rocks.

South Australia's Department of Mines and Energy knew Western Mining's secret, but was not able to tell the company's competitors to look deeper. It asked Western Mining to display publicly some core samples from Roxby Downs. The company obliged in May 1979, safe in the knowledge it already pegged all of the most prospective ground.

Western Mining's radical break from conventional thinking happened because it was unique among mineral explorers, and even among businesses the world over. In the modern language of the corporate world, Western Mining was an innovator — a Google or an Apple of its time. While it was not in the business of inventing computers and software, the mineral explorer's leaps of creative thinking were just as spectacular as those behind today's stars of innovation.

Olympic Dam was the third discovery by Western Mining in 10 years of a type of orebody entirely new to the world's mining industry. The company's exploration breakthroughs were re-writing the textbooks on economic geology.

Innovation was built into the company from the day it was founded in 1933 by W.S. Robinson, a former business editor of *The Age* turned stockbroker. He made his first fortune by sponsoring an inventor in Melbourne with an idea to extract discarded zinc from the slag heaps of Broken Hill. Robinson became hooked on the possibilities to create wealth through science and innovation, and decreed that this principle would be the foundation of his newest company, Western Mining Corporation.

The company began by recruiting the best geological brains it could find, mostly from North America, and began exploration with some strange methods that were mocked by its peers. Western Mining soon earned the name of the 'Wasting Money Corporation' because it hired planes to take aerial photos across entire regions that attracted its interest. This practice would become a fundamental exploration method by the 1950s, but was extremely unorthodox at a time when most mining companies did not even have geologists on staff. In those times, exploration consisted of scratching away at an existing orebody until an extension could — or could not — be found.

Western Mining made some great discoveries in the 1950s, 1960s and early 1970s by defying conventional thinking and going wherever science and the imaginations of its geologists told it to explore. But one target still eluded the company — copper, a highly conductive metal in demand by an increasingly electrified world.

The search had taken Western Mining's geologists to wherever prospectors had found copper minerals in any quantity, including some of the most remote locations in Australia. In the early 1970s, it tried an entirely new approach. The search would start again, but this time it would be based on theories rather than surface signs of copper.

And so Western Mining's small team of geoscientists was guided to Olympic Dam purely by their scientific ideas. They had developed some world-first thinking about where to search for copper in some of the oldest rocks in the world. Many of their theories were untested; some were even ridiculed as the stuff of mad men, but they believed in the science they had pioneered, and Parbo and his fellow directors believed in the company's scientists. It was a remarkable combination of faith at the individual level and corporate level, the likes of which may never be seen again in the resources industry.

The brilliance of the Olympic Dam discovery is enhanced with time. The world's biggest miners had all departed the region without success by the end of the 1980s. In fact, it would be 20 years after the Olympic Dam discovery before another economic find of copper was made at Prominent Hill.

Of course, organisations are only the sum of the people who work inside them. This story is really about the people who discovered Olympic Dam. At the top were great leaders like Sir Lindesay Clark, Sir Arvi, Roy Woodall and Jim Lalor, who shared a faith in geological science, trusted their people and built highly talented teams.

Woodall scoured Australia's universities for the most brilliant young geoscientists before they even graduated, and implemented a host of what today's human resource experts would call innovative talent management programs.

Western Mining's leaders were also special in the way they pursued goals with the kind of persistence that cannot be imagined in the modern corporate world. They began the search for copper in 1957 and never wavered, despite decades of disappointments and setbacks even before the start of the search for Olympic Dam.

Perhaps it was a different era, but this determination might also be specific to the individuals involved. Woodall was born in Perth in 1930 and grew up in the Great Depression. He left secondary school after three years to help support his family, finishing his schooling by evening classes. He was fascinated with science, but equally with the idea that enough wealth to sustain entire cities lay hidden below the ground. It could have great value for Australia or absolutely none if its whereabouts remained unknown.

Sir Arvi also knew about overcoming adversity. He was born in 1926 on a dairy farm in Estonia, just across the Gulf of Finland from Europe's Nordic countries. Towards the end of World War II, he fled the Soviet re-occupation of Estonia and found himself alone in Germany as an 18-year-old refugee. He would be largely separated from his family until the 1989 collapse of the Soviet Union, 40 years after he arrived in Australia to begin work by quarrying rocks in a southern Adelaide suburb.

Looking back from here, it's easy to think the sheer size of Olympic Dam made its discovery inevitable. But it is sobering to realise other Olympic Dams might still lie beneath the South Australian outback, their existence unknown because the nearest exploration hole glided by just a few metres away. In fact, there could so easily be no Olympic Dam story, but for the devotion to science, persistence and teamwork of the people in the following chapters.

How big?

The world began to see the real size of Olympic Dam in late 2007, when new owner BHP Billiton revealed the first results of a massive drilling program begun almost as soon as it acquired the mine in 2005.

The company announced a 77% increase in the size of the Olympic Dam mineral resource to 7.7 billion tonnes. To put this into context, BHP Billiton's managing director, Marius Kloppers, said Olympic Dam was now the second-largest mineral resource in the world, behind only Norilsk's nickel-copper-platinum-palladium ore fields in Siberia.

In annual reports since 2007, BHP Billiton has quietly revealed further increases in the size of Olympic Dam after more drilling to improve confidence in the size and continuity of the orebody.

By late 2009, the mineral resource had reached 9.1 billion tonnes — more than double the 4.4 billion tonnes BHP Billiton inherited from Western Mining in 2005. Olympic Dam is now also more than four times larger than the first estimate announced by Western Mining back in 1982, despite the extraction of 2.7 million tonnes of copper, 55,000 tonnes of uranium oxide, 1.2 million ounces of gold and almost 12 million ounces of silver since mining began in 1988.

When the first estimate was announced in 1982, miners, stockbrokers and other industry watchers gasped at the news that Olympic Dam was a two billion tonne resource and contained four times more copper than Australia's largest existing copper mine at Mt Isa, Queensland. So what does it mean to have a resource that is now 9.1 billion tonnes?

In volume terms, the rocks of mineable grade defined so far could fill the Melbourne Cricket Ground almost 2,000 times. In value terms, the minerals in the ground at Olympic Dam now have an estimated value of $US863,000,000,000 or $US863 billion. This includes $US470 billion in copper, almost $US270 billion in uranium, $US116 billion in gold and $US8 billion in silver.

Based on analysis of annual reports published since Kloppers made his statement in 2007, Olympic Dam has surpassed even Norilsk to become the world's most valuable mineral resource. Norilsk has an estimated, in-ground value of $US745 billion, and that's including both major ore fields (also known as camps) — Norilsk 1, discovered in 1925, and the Talnakh Nickel Camp, found 20 kilometres away across a river in 1962.

Olympic Dam is also more valuable than Andina ($US631 billion), the largest of three huge copper deposits owned by the Chilean Government through Codelco. Each of these deposits (Andina, El Teniente and Chuquicamata) contain more copper than Olympic Dam, but none have the additional value of uranium and gold.

In public presentations, BHP Billiton likes to break down Olympic Dam into its main components — copper, uranium and gold — and rank these individually against other major ore bodies of the world. This exercise shows Olympic Dam is the fourth largest copper resource in the world, the fourth largest gold resource and the largest body of uranium minerals.

Resource categories explained

The term 'total resource' is a clearly defined measure that must be used only to describe rocks that have a reasonable prospect of being mined. It excludes rocks with less than a certain percentage of metal content (grade) that would not be worth mining under any circumstances. The total resource is comprised of three categories — 'measured resource', 'indicated resource' and 'inferred resource'. These terms also have clear definitions, based on how much is known from close-spaced drilling about the location, size and continuity of the mineralised rocks. As more information is gathered from tighter spaced drilling, the status of the resource is upgraded from 'inferred' to 'indicated' and finally into the 'measured' category. Indicated and measured resources become probable and proven ore reserves respectively once the owner of the minerals can show an economically viable plan for mining the deposit. We are now finally in the territory where we can use the term 'orebody'. BHP Billiton has calculated a total resource figure at Olympic Dam of 9.1 billion tonnes. This is comprised of measured resources of 1.3 billion tonnes, indicated resources of 4.6 billion tonnes and inferred resources of 3.2 billion tonnes. The Australian mining industry began developing definitions for measuring and reporting the size of orebodies in the late 1960s to make it a safer place for investors. These standards are known as the Australasian Joint Ore Reserves Committee or JORC Code, and have been adopted by many other countries.

Olympic Dam versus the world's other greatest mineral resources[1]

	Olympic Dam, Australia		Talnakh - Norilsk 1 Russia		Andina Chile	
	Weight	Value	Weight	Value	Weight	Value
Copper (million tonnes)	78.3	470	32.1	193	105.2	631
Uranium oxide (000 tonnes)	2,441	269	-	-	-	-
Nickel (million tonnes)	-	-	16.9	313	-	-
Gold (million ounces)	96.8	116	15.5	19	-	-
Palladium (million ounces)	-	-	264	119	-	-
Platinum (million ounces)	-	-	72.0	101	-	-
Silver (million ounces)	437.3	8	-	-	-	-
Total value ($US billion)		**863**		**745**		**631**

The amount of uranium in Olympic Dam is perhaps the most staggering number. It approaches 10 times the size of the next largest uranium resource and contains more uranium than the combined total of the rest of the top 10. If the world gave up coal-fired power for nuclear electricity, Australia could command global energy supply in the same way Saudi Arabia dominates oil.

The statement that Olympic Dam is the world's most valuable mineral resource requires some qualification. It is the most valuable resource in a single location, but has a lower value than some ore fields, which are groups of geologically related mineral resources that can stretch over tens or even hundreds of kilometres. Such fields can support several mines. For example, the gold reefs of the Witwatersrand in South Africa support more than 150 mines, spread over several hundred kilometres.

Richard Schodde, a mineral economics expert and managing director of MinEx Consulting, estimates the in-ground value of the current resources of the Witwatersrand gold field to be $US1,440 billion, much greater than Olympic Dam. "The Bowen Basin coal field in Queensland is also arguably a single deposit, and collectively contains resources with an in-ground value of more than $US2,000 billion, based on prevailing coal prices. Both these examples are large and often discontinuous resources over great

---

[1] Total resource figures, which include proved and probable reserves. Based on prices for copper at $US6,000 per tonne, uranium oxide at $US50 per pound, nickel at $US18,500 per tonne, gold at $US1,200 per ounce, palladium at $US450 per ounce, platinum at $US1,400 per ounce and silver at $US18 per ounce. Andina data sourced from Codelco's 2009 annual report, page 40. Norilsk data sourced from Norilsk Nickel's 2009 annual report, page 42, and excludes minor platinum group metals. Olympic Dam data sourced from BHP Billiton's 2009 annual report, pages 70 and 71.

distances. As a single-location mineral resource, Olympic Dam is certainly out in front based on current in-ground resources."

Schodde says it's also important to take into account historic production. "If you add back El Teniente's historical production since 1910 to existing reserves, you will find it is bigger than Olympic Dam. This is also true of the Chuquicamata copper mine and the Norilsk nickel-copper operation, which have been in production since 1915 and 1942 respectively. But all of these mines have a long head-start on Olympic Dam, which is still in its youthful phase. There is every expectation reserves will continue to grow — especially at Olympic Dam."

The in-ground value of minerals is a useful measure of the size of a deposit, but more important are the value of minerals that can be extracted every year and the mine life. BHP Billiton currently extracts 12 million tonnes of ore from Olympic Dam every year via Australia's largest underground mining operation. This typically produces about 200,000 tonnes of refined copper, 4,000 tonnes of uranium oxide and 100,000 ounces of refined gold.

The company's 2009 annual reports shows sales from the mine totalled $US1.3 billion in the year to 30 June 2009 and produced a profit of $US316 million.

The total resource of 9.1 billion tonnes is enough to sustain Olympic Dam at its current production rate for the next 750 years, but BHP Billiton plans a huge boost to the rate at which minerals are extracted. Under a proposed expansion, the amount of ore to be extracted every year would climb to 72 million tonnes — a six-fold increase compared to the existing rate of mining. Production of copper, uranium and gold would increase overall by a factor of about 4.5, reflecting the lower average grade of the copper, uranium and gold minerals.

Most of the ore would be extracted by an open-cut method of mining. The proposed pit would be the world's biggest, growing over 40 years to become 4.1 kilometres long, 3.5 kilometres wide and at least one kilometre deep. Underground mining would continue, although much of the existing underground mine would be excavated, including the historic Whenan Shaft.

The effect of the expansion on revenues and profits from Olympic Dam are hard to estimate, but if production increased by 4.5 times, annual revenues would grow to about $US7 billion, assuming some increase in prices from the depressed levels of the 2009 financial year.

BHP Billiton at this stage plans to operate the expanded mine for 40 years, so we can multiply annual revenues by 40 to get a figure of $US280 billion for the cumulative mineral wealth produced by Olympic Dam in the planned-for future. But that's still only a fraction of the potential value of the mine. The total resource at Olympic Dam is enough to keep the mine running for about 130 years, even at its vastly increased rate of production. When annual revenues are multiplied by 130 years, the mineral wealth that could be extracted from the known resource at Olympic Dam increases to $US900 billion.

This is gross revenue, not profits, but it is a real measure of the wealth that could be generated by the mine. The incredible orebody could produce in the order of $1,000,000,000 (one trillion dollars) over the next century, to be distributed among hundreds of companies, including those that supply the heavy earthmoving equipment and the trucks to mine the ore, local contractors and service providers in the region, tens of thousands of BHP Billiton employees and hundreds of thousands of BHP Billiton shareholders.

The mine will also make a huge direct contribution to public revenue. BHP Billiton will pay company tax to the Federal Government on profits earned from the mine, in addition to a royalty paid to the South Australian Government. Over most of the mine's life to date, the royalty has been set at 3.5% of the value of minerals produced. This has already generated almost $500 million for South Australian taxpayers in nominal terms, and an estimated $700 million in real or inflation-adjusted terms since mining began in 1988.

None of this wealth was known until Western Mining discovered Olympic Dam in 1976. The minerals were always there, but without knowing their location they effectively did not exist as far as humankind was concerned.

Century-long estimates of any endeavour are rarely meaningful, but in the case of Olympic Dam there are good reasons to believe it will live up to its promise of being a trillion dollar mine over the next 100 years. It has the reserves. Even if they are not yet in the measured resource category, proving up ore to date has been a simple function of spending money on close-spaced exploration drilling. BHP Billiton's efforts have doubled the total resource in the past four years. Further drilling is hard to justify to shareholders when BHP Billiton already has enough proven ore to sustain the mine for decades, even at its expanded production rate. Western Mining was in the same position, with enough proven ore to support mining for decades when BHP Billiton acquired the company in 2005.

The other assumption behind the idea of a trillion dollar mine is that prices for copper and uranium do not fall to a level that makes Olympic Dam uneconomic to mine. While no-one can be confident of world demand for copper and uranium over the next 100 years, Olympic Dam has some big advantages on its side. The proposed switch to open cut mining could make it one of the world's lowest cost producers of copper and uranium over a long timeframe. This means a drop in prices would shut down the mine's higher cost competitors before Olympic Dam might be forced to follow. It should be in a position to ride through the boom and bust cycles that are a notorious part of the mining industry.

The very long-term future of Olympic Dam also looks bright because of a forecast global shift to uranium as a fuel for generating electricity. The world currently has 439 nuclear power stations, with another 36 under construction, another 97 planned and proposals to build a further 221 in the next 20 years, according to the World Nuclear Association. Global demand for uranium will double on this basis.

Given that nuclear power stations are built to operate with lifespans of up to 60 years, global demand for uranium is now being locked in over a very long time frame. Olympic Dam will be the world's largest uranium mine and a highly sought-after supplier.

The trillion dollars of wealth already known at Olympic Dam is still nowhere near the whole story. Recent drilling has shown the orebody is still open in horizontal directions and at depth, with talk in the industry that drilling as deep as 1,800 metres is still in mineralisation. In the nomenclature of the mining world, Olympic Dam is a super-giant orebody. It could be supplying the world with copper, uranium, gold and silver for several centuries, not just the next 100 years.

There is also a real possibility that Olympic Dam is part of an ore field, with BHP Billiton sitting on look-alike deposits in the immediate area, particularly at a place called Wirrda Well, only 25 kilometres south of Olympic Dam.

Douglas Haynes, whose PhD provided Western Mining with the rationale to explore for copper near Andamooka, says there is at least one Olympic Dam look-alike at Wirrda Well. "Exploration drilling by Western Mining at Wirrda Well discovered mineralisation within rocks that were closer to what we found at Olympic Dam than anything else in the region. The copper grades were similar, but this early drilling did not define the limits of the prospect because there was so much exploration focus on Olympic Dam."

Elsewhere in the region, exploration at another stock watering hole on Roxby Downs station, Appendicitis Dam, initially provided enough encouragement for Parbo to ask his geologists to rename it Acropolis. "I feared we might end up with the Appendicitis Mine, which I didn't believe to be a very fit name." Subsequent drilling revealed the rocks at Acropolis were quite different to those at Olympic Dam, with less copper and at even greater depths, but it encouraged the search for Olympic Dam look-alikes in the region.

How did it get there?

How did such incredible mineral wealth come to exist in South Australia's far north? The origin of Olympic Dam is still hotly debated, despite 35 years of investigation, 2.3 million metres of diamond drill core and 400 kilometres of underground mine development.

The origins of many orebodies are not well understood, which seems remarkable given their importance to the global economy. Western Mining's former chief geologist and director of exploration, Roy Woodall, says "we're not even at first base" in understanding how some orebodies are formed. "For example, porphyry copper deposits are perhaps the most uniform of all types of ore body and have been studied very extensively, but there is much we still don't know about how they were formed."

Douglas Haynes says although there are still many puzzling questions about the genesis of large orebodies, particularly the sources of metals in some styles of deposit, much progress has been made. "We are forming a good understanding of the genesis of the better-studied orebodies, such as large copper deposits in higher-level igneous intrusions known as porphyries, and for certain types of gold deposits, particularly those where the gold is in quartz veins. But among all orebodies, Olympic Dam is one of the most difficult to unravel because of its unusual host rocks, metal associations, geological setting and lack of exposure at the surface."

Let's start with the few things that are not debated about Olympic Dam and its geological setting. First, it's agreed that Olympic Dam is unique. It was the first ore body of its kind to be discovered, with a never-before-seen combination of iron oxide, copper, uranium and gold. After the discovery in the 1970s, geologists worldwide re-examined existing deposits to see whether Olympic Dam was truly a one-off. They found enough common characteristics with some other deposits, mostly in Chile, to devise a new category of orebody known as Iron-Oxide-Copper-Gold or IOCG. But it's agreed that Olympic Dam is still on its own in many respects as a copper deposit, especially in regard to its uranium content and high content of elements known as rare earths.

Second, we know that Olympic Dam is within a giant complex of broken or brecciated rocks that covers an area of about 50 square kilometres. The orebody itself occupies a central area of about 20 square kilometres. The breccia is a geological fruit salad of almost every major rock type. It includes fragments of Roxby Downs Granite, which is a reddish-coloured igneous intrusive rock, and fragments of other igneous intrusive rocks. These include dolerite and unusual rocks known as ultramafics. The breccia also includes fragments of a mixture of volcanic and sedimentary rocks. The entire deposit is concealed by 350 metres of flat-lying sediments, deposited between 1,000 million and 550 million years ago.

Third, we know Olympic Dam is within an ancient chunk of continental crust known as a craton (pronounced 'kray-ton' and derived from the Greek word for strength). Olympic Dam is near the northeastern edge of the Gawler Craton. Cratons are the oldest building blocks of continents and have survived many cycles of continent building and destruction. They extend to depths of 50 kilometres or more, making them thicker than most parts of the Earth's crust and far more durable. A branch of geoscience has devised techniques for accurately tracking the historical movement of cratons across the Earth's surface under the force of plate tectonics. The change in latitude of cratons with time can be traced from the changing imprint of the Earth's magnetic field on ancient basalts as they cooled. It shows cratons have travelled half-way around the world and back again, often a number of times.

Every continent today has cratons around which new landmass has formed. Australia has four main cratons — the Gawler in South Australia, the Yilgarn and Pilbara Cratons in Western Australia and the Northern Australian Craton that extends across the North Territory and into western Queensland.

Cratons are rich sources of economic minerals because of their very old age and involvement in the formation and break-up of many continents. They are more likely than most young rocks to have been subjected to one of the many geological processes that can concentrate widely dispersed minerals into an orebody.

Igneous basics

Igneous rocks are formed from cooling of molten magma into solids. When magma erupts at the surface it is known as lava. It cools quickly and the molten minerals do not have time to grow into large crystals. The result is a fine-grained rock known as a volcanic or extrusive igneous rock. Basalt is one of the best examples. When magma does not reach the surface and cools slowly underground, larger crystals grow and the rock is coarse grained. The resulting rock is known as an intrusive igneous rock, with granite one of the most common examples.

The Gawler Craton covers three-fifths of South Australia, but is almost entirely hidden by younger rocks. Only five percent of it is exposed at the surface. Geoscientists have even looked for clues about Olympic Dam in Antarctica, which was connected to the Gawler Craton until about 250 million years ago — quite recent in terms of geological time. But in Antarctica, the thick cover of ice has replaced the parched sediments of South Australia's outback as a barrier to understanding the ancient processes that formed the orebody.

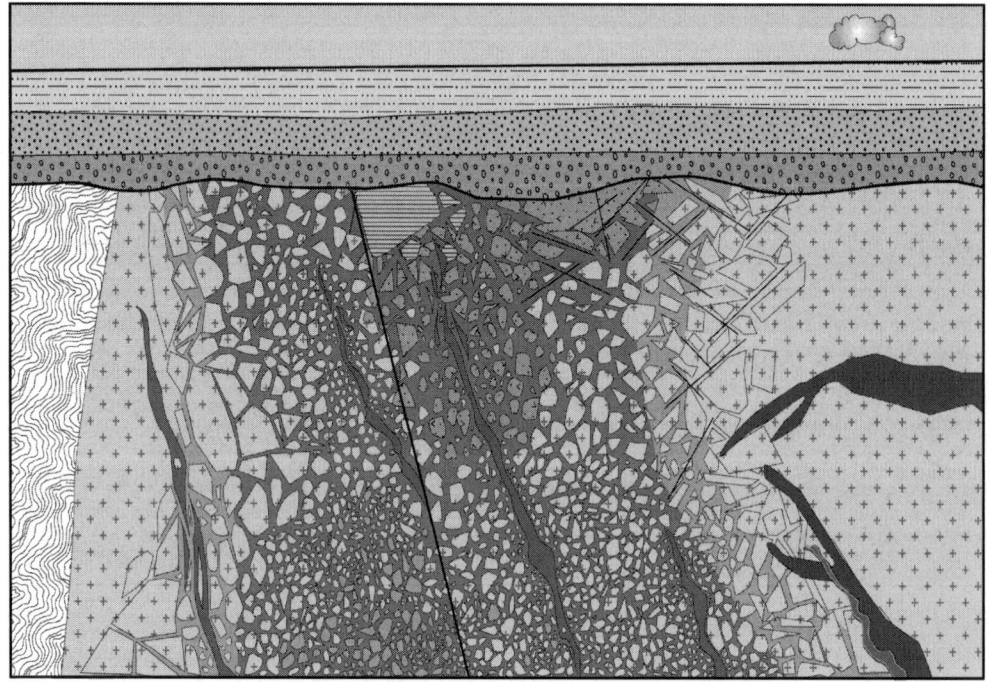

This graphical cross-section of the Olympic Dam orebody is from an April 2010 presentation by BHP Billiton Uranium's Principal Geometallurgist, Kathy Ehrig to the 13th Quadrennial Symposium of the International Association on the Genesis of Ore Deposits. It incorporates new information from extensive drilling between 2005 and 2008. It is the most complete view of the orebody ever released and shows the complexity of the Olympic Dam resource. The flat-lying sediments above the orebody are more than 300 metres thick. The graphic shows only the top 1,200 metres of the orebody, which continues to unknown depth.

The final bit of information that's largely agreed about Olympic Dam is the age of the deposit at 1,590 million years, although some research shows that parts of the orebody contain much younger components, such as some mineral veins. The most reliable estimates of the age of the deposit were determined by radiometric dating of zircon crystals in the deposit. While there are older orebodies in Australia, such as the great iron ore deposits of the Pilbara at about 2,500 million years old, few are in settings where there have been so many major geological events. These have over-printed the geological record several times and made it extremely difficult to unravel the history of the Olympic Dam deposit.

So how might it have formed? The prevailing view — developed in the 1990s by Western Mining geologists Ken Cross, Jim Reeve and Douglas Haynes — is the Olympic Dam deposit formed in a hot, steaming volcanic lake, between 300 and 1,000 metres below the surrounding landscape. It was probably in a rift valley formed by the pulling apart of tectonic plates, as seen today in the northern African Rift Valley in Ethiopia.

This was no ordinary volcano that grew tall, became dormant and gradually wore away over millions of years. The Olympic Dam volcano blew itself to pieces at regular intervals when water in the crater drained suddenly onto hot magma. Geoscientists call this a phreatomagmatic eruption. It's an even larger and more powerful form of the eruption that is believed to have destroyed the Indonesian island of Krakatoa in 1883, leaving only a small remnant of an island that was formerly 22 kilometres long and more than 800 metres high. Krakatoa was a phreatic eruption, in which gas is released with explosive force from the mixing of different kinds of very hot magma. A phreatomagmatic eruption is even more powerful because it is caused by the sudden mixing of huge volumes of two liquids of extremely different temperature — water and magma. Steam is released instantly with unimaginable explosive power. These eruptions are characterised by massive volumes of broken rocks rather than outpourings of lava.

The landform that remains after a phreatomagmatic eruption is known as a maar — a low-relief volcanic crater that fills with more water and forms a hot, steaming volcanic lake. In the geologically brief period of time of a few thousand years, the conditions are set for another gigantic explosion as another pulse of magma enters the deep, water-filled conduit of the volcano beneath the lake. Maar volcanoes are known to be involved in the origin of some of the largest mineral deposits in the world, including the Grasberg copper-gold mine in Papua New Guinea and the spectacularly rich Cripple Creek gold deposit in Colorado, USA. Maar volcanoes are sites of extreme variation in temperature, pressure and geochemistry, making them breeding grounds for ore-forming processes.

There is ample evidence of a maar volcano at Olympic Dam. Most of the economic minerals are within a near-vertical zone that might have fed magma to an ancient volcano from deep in the Earth. The zone is filled with breccia or broken fragments of granite and other rocks that were likely created by hundreds of phreatomagmatic and lesser phreatic explosions. There's also evidence that Olympic Dam's copper, uranium, gold, silver and iron minerals could only be deposited by the mixing of two extremely different fluids — a hot fluid from deep below the volcano and another in the form of lakewater and groundwater in and around the crater.

The maar volcano theory is now being challenged by researchers studying the data from about 600,000 metres of recent diamond drilling by BHP Billiton. New hypotheses for Olympic Dam include the idea that fluids circulating through sediments overlying the deposit played a part in its formation. This is a radical suggestion because it says the economic minerals in Olympic Dam were deposited perhaps as recently as 600 million years ago — a billion years later than the maar volcano theory. It's not the first time a much younger age has been suggested for the orebody; researchers at the University of Arizona put forward the same idea in the 1980s, based on the sampling of younger mineral veins. It is strongly rejected by Cross, Haynes and others, who argue the hypothesis is based on sampling that is not representative of the orebody. Advocates of a much younger age for Olympic Dam argue there are too many sedimentary rocks associated with Olympic Dam — more than previously recognised — to be consistent with the frequent and violent eruptions of a maar-volcano setting.

The source of the copper, uranium and gold is also debated. The copper has two or three potential sources, depending on the model of Olympic Dam's genesis to which you subscribe. The most established model says some of the copper leached from basalts overlying the surface of the Gawler Craton, and some was introduced from the hot fluids below. The basalts were largely eroded away before the deposition of overlying sediments. Erosive forces would have been fierce because high concentrations of fluorine were present and would have dissolved in groundwater to create hydrofluoric acid. Another leading theory for the source of the copper and gold points to the large sheets of originally magnesium-rich igneous rocks known as ultramafic dykes that cut across the breccia complex after it began to form. The most accepted view is the copper and gold were sourced from the basalts and the dykes, which were part of the same igneous event.

There is less contention about the source of the uranium in Olympic Dam, which is almost an order of magnitude larger than the world's next-biggest uranium deposit. It's largely accepted the uranium was brought to the surface by rocks known as the Roxby Downs Granite or its pale-coloured volcanic equivalents, known as rhyolite and dacite. At the time Olympic Dam was formed, the Gawler Craton experienced a massive amount of igneous activity over a geologically brief period of time. Huge volumes of lava poured over the ancient landscape to create the Gawler Range Volcanics, while magma that did not breach the surface cooled slowly to become an associated rock formation known as the Hiltaba Suite Granites. Combined, the Gawler Range Volcanics and the Hiltaba Suite Granites, including the Roxby Downs Granite that hosts Olympic Dam, have a volume of about 100,000 cubic kilometres. If this volume of igneous rock could be cut into inch-thick slabs, it could pave the total land surface of the world more than 25 times. It represents one of the biggest magmatic events in the Earth's history.

What made it so hot below South Australia's far north about 1,600 million years ago? Once again, there are several theories and no certainty. One idea suggests Olympic Dam was part of a large arc of volcanoes created by the collision of two tectonic plates. Volcanic arcs lie around the Pacific Ocean today, where oceanic crust collides with continents and is subducted deep into the Earth.

Geoscientists agree Olympic Dam is near the edge of an ancient piece of continental crust, and there's some support for the idea this boundary had another plate slamming into it at this time. The second plate would not have been part of eastern Australia as we know it today, which did not exist until about 500 million years ago. But perhaps another ancient chunk of continent — the nucleus of North or South America — may have collided with the Gawler Craton at this time. This theory has problems, however. The granites and volcanic rocks of the Gawler Craton are not a good fit with the composition of rocks geoscientists normally see from the collision of two continental plates. Furthermore, South Australia has a very old and complex geological history and it's hard to know whether there was any plate collision at this time. The state's complicated coastline of jagged, pointy gulfs and the vast, low salt lakes in the state's interior give the impression that South Australia is trying to split in two. Geologists believe the interesting geography of the state reflects a rift that began but then failed to continue. But all these events took place less than 800 million years ago, barely half the age of Olympic Dam. Even the oldest parts of the Flinders Ranges, the state's most famous geological feature, go back no more than 1,000 million years and are the product of separate and much younger events. There are many complicated chapters of geological history extending way beyond the time we can reasonably understand.

A currently favoured theory is that a mantle plume was responsible for the massive volcanic activity in the Gawler Craton when Olympic Dam was formed. Mantle plumes are unusual and mysterious upwellings of molten rock from the deepest part of the Earth's mantle, starting at about 3,000 kilometres below the surface. The chemical composition of the Earth at these depths is very different to the crust beneath cratons and other continental rocks. Molten rock that makes it to the surface from these great depths typically appears as basalt, an iron and magnesium-rich volcanic rock. The basaltic rocks from the Mantle plumes are so hot — about 1,600 degrees Celsius — they melt parts of the old craton, producing other types of igneous rocks such as granites, rhyolites and dacites.

Mantle plumes are largely accepted as the reason for volcanic activity away from the edges of tectonic plates. Like an oil bubble in a lava lamp, the plume travels upwards and begins to develop a broad, flat head as it nears the top. The plume rises at a rate of about 100 kilometres every million years, which means it takes a geologically brief time period of 30 to 40 million years to travel from the bottom of the mantle to the Earth's crust.

The volcanic island chain of Hawaii is regarded as the most famous example of a mantle plume. It occurs within a region of oceanic crust, which is very different from the continental crust of the Gawler Craton. The islands in the chain are progressively older towards the northwest. Mantle plume theory says this is because the plume upwells at a fixed point in the mantle, and the overlying tectonic plate is travelling in a northwest direction across the top.

Research suggests that a major mantle plume 1,590 million years ago stretched part of the continental craton at Olympic Dam, causing it to crack and form rift valleys as the plume rose. It then heated the crust, partially melting it to form massive and extensive granites and the pale coloured igneous rocks. Associated with this event were basalts and basaltic dykes where parts of the plume head reached the surface through the heated crust.

Recent drilling at Olympic Dam has confirmed large amounts of remnant basalt overlying the ancient land surface of the Gawler Craton.

Haynes points out that basalts of the correct age and chemical composition for a mantle plume origin are seen to outcrop about 350 kilometres south of Olympic Dam. These basalts are known as the Roopena Volcanics and played a critical role in guiding Western Mining to Olympic Dam. In 1974, the company's geologists discovered the Roopena Volcanics are vastly depleted in copper minerals. While this might seem to be disappointing news for a copper exploration team, the Roopena Volcanics caused great excitement; they were exactly what the innovative Western Mining team had hoped to find after an Australia-wide search over the past two years.

Plate tectonics, cratons and recycled continents

Plate tectonics says the surface of the Earth is a jigsaw puzzle of hard plates, floating on top of fluid-like layers in the Earth's mantle. These plates are huge — only eight major plates circle the earth, and have an average thickness of 200 kilometres beneath continents and about 100 kilometres beneath oceans.

The plates are spreading apart from mid-ocean ridges and colliding together at other boundaries. The forces driving tectonic plates are not well understood, but plate tectonic theory has been formally accepted by the world's geoscientific community since 1984. You don't need to be scientist to get the idea. Many people notice South America and Africa look like adjacent pieces of a jigsaw puzzle, divided by the Atlantic Ocean. In fact, the geological evidence says these landmasses were joined about 250 million years ago. All the continents of the world were clustered together at this time to form the Pangaea supercontinent, surrounded by a single ocean. But 250 million years ago is recent in geological time, which extends back beyond 4,000 million years. If today's continents could move into their current positions in such a relatively short period of time, imagine how much else might have happened in the 1,600 million years since Olympic Dam was formed?

Pangaea was only the latest in a series of supercontinents that have come and gone thanks to the relentless forces of plate tectonics. Pangaea was preceded by Pannotia, which survived from 600 million to 540 million years ago. One of the components of Pannotia was Gondwana, a large continental block made up of Australia and Antarctica. The cycle before Pannotia created the supercontinent of Rodinia, which existed between 1,300 million and 750 million years ago.

Continents not only move, but undergo dramatic changes of their own. Vast areas are added by volcanic activity; crust can be pushed into high mountains; erosion wears away volcanoes and mountains and re-deposits them as sediments over even wider areas. Parts of continents can also disappear, subducted into the Earth where they vanish with almost no trace. For scientists trying to unravel these events, it is like trying to put together a jigsaw puzzle with pieces that every so often get chopped up and put back together in different shapes. It's hard to get a clear picture of the world at the time of Rodinia, and it becomes even more difficult if we go back even further to the time Olympic Dam was formed.

4

DRIVEN TO EXPLORE

If any company was going to find Olympic Dam, it was Western Mining. The Melbourne-based resources company was unlike any other in Australia — and possibly the world — because of its extraordinary willingness to invest in exploration. Through the 1960s and 1970s, it spent an average of 56%[2] of its annual profits on exploration. To put that into context, Western Mining's annual exploration budgets in absolute dollar terms matched those of Conzinc Rio Tinto and BHP, yet these companies were three-times and six-times bigger than Western Mining respectively. This is one of many surprising facts in a 1975 report[3] by McKinsey & Co., commissioned by Conzinc Rio Tinto in a bid to crack the secret code of Western Mining's exploration success.

Comparisons with industrial companies put Western Mining in a broader perspective. Industrial companies invest in research and development (R&D) in the same way miners spend on exploration. Apple Inc., which has earned a reputation as the world's most innovative company, spent 23% of its net profit on R&D in 2009, less than half the proportion of profits committed to exploration by Western Mining through the 1960s and 1970s.

For all its spending on exploration, Western Mining was not financially reckless. Shareholders were patient and for the most part content to see their company's profits re-invested in exploration rather then returned to them as dividends. They had seen the company make great discoveries and been richly rewarded by a soaring share price. The discovery of nickel at Kambalda in 1966 boosted Western Mining's share price by a factor of 10. If Western Mining wanted to spend nearly all its profits on exploration — and the discoveries kept coming — that was fine with shareholders.

---

[2] Western Mining annual accounts (from an article by Robert Gottliebsen in *The Australian Financial Review*, 3 August 1979).

[3] *Successful Management of Minerals Exploration in Australia*, McKinsey & Co, June 1975

Western Mining net profit and exploration spending in 2009 dollars

Year	Net profit ($ million)	Exploration spend ($ million)	Exploration spend/net profit (%)
1935	-0.8	13.5	-
1936	-0.3	2.2	-
1937	1.0	0.3	30
1938	3.1	1.9	61
1939	3.1	1.6	52
1940	1.9	1.6	84
1941	0.5	0.2	40
1942	1.2	0.7	58
1943	4.0	0	0
1944	5.8	0	0
1945	2.8	0	0
1946	1.8	0.2	11
1947	5.1	5.8	114
1948	4.8	7.0	146
1949	12.6	1.4	11
1950	6.6	1.3	20
1951	3.6	0.3	8
1952	2.3	0.05	2
1953	5.5	0.08	1
1954	7.2	0.2	3
1955	6.1	1.1	18
1956	7.8	0.3	4
1957	6.3	0.15	2
1958	6.7	0.4	6
1959	7.8	1.3	17
1960	6.2	1.4	23
1961	7.1	2.9	41
1962	6.4	6.2	97
1963	6.6	5.6	85
1964	7.3	7.2	99
1965	10.8	10.3	95
1966	8.7	5.9	68
1967	9.5	6.0	63
1968	23.1	15.4	67
1969	26.3	21.8	83
1970	136.5	27.1	20
1971	193.8	29.0	15
1972	110.9	26.7	24
1973	129.0	21.0	16
1974	97.3	27.9	29
1975	84.4	30.4	36
1976	60.9	25.1	41
1977	101.7	38.2	38
1978	43.0	54.3	126

Western Mining was special in one other way; it was determined to try new exploration ideas and technologies, even when its competitors sometimes laughed at it. In the absence of such boldness and determination, it is possible to believe that Olympic Dam — a wild idea even by Western Mining's standards — might still be undiscovered.

Western Mining's aggressive approach to exploration was chosen very purposefully by William Sydney Robinson when he founded the company in 1933. Even more remarkable was the fact Western Mining's leaders stuck to this boldly different path for most of the company's 72-year history. The sense of corporate purpose was incredibly strong, thanks in large part to having only three chief executives after Robinson in its first 57 years, all of whom were mining engineers.

Robinson was effectively the first chief executive, although his vast spread of interests around the world allowed only limited involvement with the company. Knowing this, he appointed at the outset a young Tasmanian mining engineer, Gordon Lindesay Clark (later Sir Lindesay) as Technical Managing Director.

Lindesay began working part-time for Gold Mines of Australia in 1930, going into the field to look at gold exploration proposals offered to the new company. He retired as the top executive of Western Mining in 1962, but continued as chairman until 1974. His influence on the company spanned 41 years.

When Robinson founded WMC, he was aged 57 and already a great mining and industrial leader. He had come a long way from his humble beginnings as a fruit farmer on his brother's orchard at Ardmona, Victoria. He left the fruit farm in 1896 to join his father as a business journalist at *The Age*. Robinson had the distinction of being the newspaper's first journalist to use a typewriter. He hired a Remington machine in expectation of getting a job translating his father's handwriting into something the typesetters could actually read. He got the job, although he was initially spending one quarter of his wages just to hire the typewriter. It was an early sign of his lifelong passion for new technology.

In a few years, Robinson succeeded his father as business editor and even became a fortnightly contributor to *The Times* and *The Economist* in London. But after 10 years in journalism, he accepted an invitation to join Lionel Robinson, Clark & Co., the stockbroking firm established by his eldest brother, Lionel. The firm had thrived on the mining boom created by the discovery in the 1880s of lead, zinc and silver at Broken Hill, NSW, and the discovery of gold at Kalgoorlie, WA, in the 1890s.

In 1899, Lionel Robinson moved to London and bought a seat on the London Stock Exchange. He followed a trail well worn in those days by Australian brokers and financiers. London investors were the source of most of the world's capital and hungry for opportunities to earn rich rewards from investing in new mineral discoveries around the world.

By 1905, the gold boom slowed and Lionel Robinson needed new opportunities to promote to investors in London. He believed the time might be right for a revival of mining at Broken Hill. The immensely rich orebody in the Barrier Ranges of western New South Wales was the foundation of the Broken Hill Proprietary Company Ltd, the forerunner of today's BHP Billiton. The company dominated the city and was thought to hold most of the remaining reserves of lead, zinc and silver.

Lionel Robinson travelled to Australia to organise a train journey from Melbourne to Broken Hill via Adelaide. He planned to take a party of the country's leading men in mining and finance to see first-hand what opportunities might still be available in Broken Hill. Lionel Robinson invited his younger brother to join the trip because he had been impressed by his recent writings on the possibility of rises in base metal prices.

The trip was life-changing for W.S. Robinson, as he recorded in his memoirs, entitled *If I Remember Rightly*[4], published in 1967, four years after he died. "The week's visit was a wonderful experience for me. Mountains of zinc tailings glittered in the sun, and I was told that existing treatment processes could extract much of the lead and silver but virtually none of the zinc. The challenge and rewards to anyone who could come up with a new process for extracting zinc were immense. Underground, I saw great lodes which, to my inexperienced eye, appeared never ending. As it turned out, another bumper reward awaited those men who foresaw that the lodes did in fact extend horizontally. I was impressed and showed it at every turn, so that in a very short time 'Joe' Baillieu, the broker, had me dubbed as 'Billy, the Barrier Bull'. I started writing then on Broken Hill and seem to have been writing ever since."

Robinson and the powerful friends he made on that trip, including William (W.L.) Baillieu, the founding father of Melbourne's famous family, immediately began an extraordinary revival of Broken Hill. They made their fortune by taking up the challenge of finding new technology that could unlock the zinc in the waste rocks that dominated the skyline of Broken Hill. The Zinc Corporation was formed in 1905 to develop a new process stumbled upon by a Belgian brewer, Auguste de Bavay, who worked in one of the Melbourne breweries directed by William Baillieu. The new process extracted zinc minerals by attracting them to the surface of bubbles. The minerals rose to the top of a liquid sludge of mine waste, leaving the residues of barren rock at the bottom. It was reversal of the gravity separation processes that been used for centuries to treat ores.

---

[4] Cheshire Publishing 1967, edited by Geoffrey Blainey.

A flotation plant was built at Broken Hill, a much bigger version of an experimental set-up at the Melbourne brewery. It took several years of trials to perfect the process, which caused great anxiety for William Baillieu and Robinson. They had jointly guaranteed a loan to finance the construction of the plant, and struggled to attract investors to cover the mounting costs of unsuccessful trials. In fact, they were among only a handful of believers, along with Herbert Hoover, an American mining engineer who arrived independently at Broken Hill at the same time as Robinson with similar ideas of reviving the city. Robinson and his group initially saw Hoover as a potential competitor, but soon co-operated. It was a great friendship to make with a man who would be elected President of the United States in 1929.

By 1908, the flotation process was successful. The fortunes of Zinc Corporation changed and it was soon paying fabulous dividends. In his memoirs, Robinson recounted that "no metallurgical development in the last fifty years, at least, added so much to the wealth of the world as the flotation process."

Zinc Corporation's breakthrough created the financial resources for the company to make a direct move into mining and exploration. It had made a fortune by defying conventional thinking and using innovative exploration methods. The company acquired mining leases north and south of the known Broken Hill lode, where previous explorers had mistakenly concluded the ore did not extend.

These bold plays were the basis of a powerful new family of mining companies, including Broken Hill South, North Broken Hill and Broken Hill Associated Smelters, collectively referred to as the Collins House group because of their shared headquarters at 360 Collins Street, Melbourne. Their influence continues today. Zinc Corporation became a foundation of Conzinc Rio Tinto Australia or CRA. In turn, CRA would merge in the 1990s with the Rio Tinto Company of the UK to form today's Rio Tinto group.

Robinson's fascination with technology and bold exploration plans helped turn a group of " interlopers and nobodies" into a rival to BHP. And they did it with the very resource BHP thought it controlled. These lessons would be powerful in shaping Robinson's radical template for Western Mining when he turned his attention to gold almost 25 years later.

Robinson goes for gold

Gold mining is synonymous with fabulous wealth, but it hasn't always been that way. In the 1920s, mines around Australia's gold mining capital of Kalgoorlie were closing every week. Thousands of workers were deserting the city, and local real estate could not be given away. The Great Depression of 1929 was still years away, but hard times had already hit Australia's gold mining industry because of economic decisions by the British Government after World War I.

From 1870 until the beginning of the War in 1914, Britain and other leading developed countries had adopted what was known as the international gold standard for their currencies. Governments tied the value of their currencies to gold because it was the most widely accepted and trusted store of value in the world, much more so than paper currencies. In effect, governments warranted that every unit of currency was backed by a specified amount of gold, and built up large reserves of gold bullion to reassure investors in their currencies. Leaving nothing to chance, the leading governments of the day also fixed the gold price. This allowed governments and business to invest and borrow across borders, without fear of a foreign exchange disaster.

Up to 1914, the gold price was fixed at $US8.50 per troy ounce, but the system collapsed with the outbreak of World War I. In the free-market that followed, investors clamoured for gold bullion because it was an indestructible safe haven for their wealth. Gold prices climbed as high as $US11.25 per troy ounce by 1920, but the good times for gold investors were about to come crashing down. The huge cost of the War had effectively halved the real value of pound sterling. England's Chancellor of the Exchequer, Winston Churchill, was determined to restore the pound to its former glory, even if it meant artificially over-valuing the pound against other currencies and gold bullion. In 1925, he returned England to the gold standard and the pre-War fixed gold price of $US8.50 per troy ounce. Australia was locked into this move because the Australian and English pounds were tied at parity. Local gold producers were hit with a double whammy of lower gold prices and an uncompetitive currency. The rest of the Australian economy was also affected because many of the country's biggest sources of income — wool, wheat and metals — could not be sold on world markets without substantial price cuts to compensate buyers for the over-valued Australian dollar.

By 1929, the economic situation was grim, and it was about to get much worse as the effects of the worldwide economic depression reached Australia's shores. But the fortunes of the gold industry were again to change suddenly, this time for the better. The Australian Government in 1931 devalued the local pound against sterling in an effort to save the economy from fast-approaching economic depression. The exchange rate was lowered from 1.0 to 1.25 Australian pounds to every English pound. The price of gold in Australian dollars effectively rose overnight by 25% and gold mining was suddenly profitable again.

Robinson, who now ran scores of mining ventures around the world from his base in London, knew the devaluation was coming. He formed a syndicate of investors in London with the sole aim of pursuing gold exploration and mining in Australia. In typical Robinson style, the move into gold exploration in Australia was bold. He aimed for big rewards and knew this meant taking big risks. But once again Robinson set out to use the latest scientific ideas to give this new venture the greatest chance of success.

By this time, Robinson was well connected with leading mining men from around the world. They clearly held him in high regard because they were eager investors in his planned revival of the Australian gold industry and sat on the boards of the new companies he created. The first company to be formed, in April 1930, was Gold Mines of Australia Ltd (GMA), with a registered office alongside the many other Collins House companies at 360 Collins Street, Melbourne. The main shareholder of GMA was New Consolidated Gold Fields Ltd, a famous gold mining company in South Africa. It held a 60% interest and controlled the company. The other shareholders were The Zinc Corporation Ltd, of which Robinson was managing director, the Imperial Smelting Corporation Ltd and some private shareholders.

The company's main hopes lay with two of the great historical gold producing regions of Australia — Kalgoorlie in Western Australia, where gold had been discovered 40 years earlier, and Bendigo in Victoria, where a gold rush 80 years ago had laid the economic foundations of Melbourne. However, it received exploration proposals from all states of Australia, and its first income was from a small mine in Queensland at Mt Coolon, about 900 kilometres up the coast from Brisbane, inland from Mackay.

In 1933, GMA reorganised its activities to cope with the large number of exploration opportunities and their wide geographic spread. GMA retained its interests in the east of Australia, while Western Mining Corporation was formed on 2 March 1933 to take over most of the Western Australian interests.

In July 1934, another corporate reorganisation took place — a new vehicle, Gold Exploration and Finance Company of Australia Ltd (GEFCA) was formed, with Western Mining as an almost wholly owned subsidiary. GEFCA became the holding company for what was already a sprawling group of gold exploration and mining ventures created by Robinson's busy mind. The new company's most important role was to raise the large amounts of capital necessary for the group's increasingly ambitious gold exploration plans. On 30 July 1934, GEFCA issued a prospectus in London to raise £900,000, equivalent to about $A75 million in today's value. The issue was filled the following day.

Investors were willing to provide vast sums for gold exploration because the share prices of gold companies rose strongly through the Great Depression, defying the trends in almost every other industry. The gold industry then received a major boost in 1934 when the US Government lifted the fixed price of gold from $US20.67 an ounce to $US35 an ounce as one of a number of measures to kick-start the US economy. It triggered a worldwide scramble for gold stocks, but Australia's participation in the boom was no certainty. The major gold mining centres in Australia were long past their peak and had been in decline for decades.

The genius of Robinson was to attract the world's leading miners and investors to the distant and run-down gold mines of Australia on the basis that great riches were still waiting to be discovered. He succeeded in bringing a "who's who" of investors to GEFCA, including Central Mining and Investment Corporation, at that time the largest gold producer in South Africa, and Union Corporation, also from South Africa. Another eager investor was Walter Case of Case, Pomeroy & Co. of North America.

WMC scorned as the "Wasting Money Corporation"

Western Mining's new exploration programs began in spectacular fashion. Robinson, on the advice of New Consolidated Gold Fields director, Robert Annan, hatched an ambitious plan to find undiscovered gold fields by taking aerial photographs of vast tracts around Kalgoorlie. It was originally planned to shoot up to 200,000 square kilometres — almost 10% of the area of Western Australia — over two or three years, with most of the important areas around Kalgoorlie to be completed by the end of 1934.

Robinson had already cleared the way by negotiating with the West Australian Government a new form of territorial lease, which would give Western Mining security over the new discoveries it hoped to make from the air.

Wing Commander Victor Laws, who led aerial photography for the British Expeditionary Forces in World War I, was hired to implement the plan with military precision, and no expense was spared. Two aircraft, *The Golden West* and *The Gay Prospector*, were purchased and extensively modified for their new task. The De Havilland Dragons needed a stronger undercarriage to cope with the outback airstrips and better-than-average instruments for precise navigation.

A convoy of trucks supported the survey, carrying photographic equipment, an air-conditioned darkroom and three wireless telegraphic stations. It was a technological tour de force. Sir Arvi Parbo recalls that it soon earned Western Mining the nickname of the "Wasting Money Corporation" from bemused onlookers in the Kalgoorlie gold fields.

Unfortunately for Western Mining, the cynics were right. The aerial photos did not reveal gold, but vast expanses of soil and laterite, a weathered, gravelly layer, which masked the rocks of real interest below. Australia was an older and far more weathered terrain than other countries where aerial surveys had yielded results for a new generation of gold explorers.

Robinson also learned that unsophisticated prospectors had been far more efficient over the previous 40 years than anyone expected; the most promising outcrops of gold-bearing rock located by the aerial survey had already been picked over.

The poor results from the aerial photographic surveys did not discourage Robinson. He already had underway another great experiment in which he would recruit the world's best geologists to unlock the secrets still held by Australia's goldfields. It seems an unremarkable plan today, but in the 1920s and 1930s geology was a still a new science and regarded as a poor cousin in the study of natural sciences. It had yet to make an impact on the mining industry, which relied on the discoveries of prospectors from scratching at minerals exposed at the surface.

But a revolution was underway, led by a team of geologists at Harvard University in Massachusetts, USA. Pioneering work in structural and mine geology was taking place under Professor of Economic Geology, Donald H. McLaughlin. This new knowledge had already been put to extremely good use. McLaughlin had rejuvenated the historic Homestake gold mine in South Dakota, making a fortune for Walter Case, who was now among GEFCA's new investors.

Careful geological mapping at Homestake identified for the first time the rock units in which the gold occurred, while some innovative interpretation pointed out where else it would most likely be found. Now Robinson, with the assistance of Case, wanted to bring this revolution to Australia. He seemed to have no trouble securing the help of McLaughlin, who set up Western Mining's geology department, selected its geologists and established the standards it was to follow.

McLaughlin nominated another Harvard-trained geologist, Dr Hugh McKinstry, to go to Australia to lead Western Mining's new geological team. McKinstry had trained at Harvard under McLaughlin, and was regarded as a world leader in the geology of Precambrian rocks, which hosted the gold mines of Kalgoorlie.

McKinstry had famously revived the Hollinger gold mine in Canada, another Precambrian ore body with many similarities to those found in Western Australia. In 1934, he arrived in Australia to head the team, which eventually included seven geologists from North America. Many of them would go on to rank among the famous geologists of their generation, including John Gustafson, Terrence Connolly and Haddon King.

The early going in Australia was slower than expected. There were no quick wins, as many of Western Mining's influential shareholders had hoped. A number of them withdrew their financial support, including Walter Case and Robert Annan. By 1936, Western Mining, Gold Mines of Australia and their parent company, GEFCA, were perilously close to being wound up because of a lack of funds.

But gold production was finally beginning, generating much-needed cash. Mines in the central fields of Kalgoorlie and the Triton mine, near Cue, had been brought into production. Other new mines in Western Australia — Cox's Find, near Erlistoun in the heart of the state, and at Norseman, south of Kalgoorlie, were also about to produce.

These would secure the future of Western Mining, at least for the time being. The best of science and a big commitment to exploration had indeed carved out a place for a new gold mining company in Australia. But the other lesson learned in these early years was the need for patience and persistence. This knowledge would be invaluable when the search that culminated in the discovery of Olympic Dam began in the 1950s.

Clark takes charge

Robinson had the vision to create Western Mining as an innovative explorer, but he was never as close to the company as Clark. Robinson was based in London and constantly travelled the world, looking for new ventures in many areas of enterprise, not just resources. Day-to-day management was left to Clark, the young Tasmanian mining engineer installed by Robinson on day one as the Technical Managing Director.

It should have been no surprise that Clark developed a much greater love of the company than Robinson. But Robinson would have been taken aback in the early 1950s when Clark acted like the protective father of an adopted child to keep Western Mining out of the hands of Robinson and his English associates.

The showdown was set in motion in 1949. Until then, Western Mining's ownership structure was largely unchanged from its inception. It remained an almost wholly owned subsidiary of GEFCA, which in turn was dominated by English and South African investors. By this time, the South Africans had lost interest in Western Mining, preferring to focus on richer gold exploration opportunities that were emerging at home. They were willing sellers of their GEFCA shares.

The English also were unhappy because Western Mining shares traded at higher prices on stock exchanges in Australia than in London. Australian investors valued gold shares more highly because the gold industry was tax-free. The Australian Government had a long-standing policy to exempt the industry from taxes, in recognition of its role in opening up regional areas and driving the development of railways and other infrastructure.

Robinson set about a major reorganisation of the group. Western Mining took over its parent GEFCA, and new Australian shareholders were introduced to effectively replace the South African and English investors. Robinson resigned as a director of GEFCA as part of the 1949 restructure. This was the last office he would hold in connection with Western Mining.

At the same time, Clark's influence over the company was strengthening. In 1952, he became managing director and chairman, following the death of previous chairman Sir Walter Massey-Greene. Western Mining was now an Australian company, with its destiny in the hands of local directors and management, and an increasingly large and loyal group of Australian shareholders. Clark planned to keep it that way.

Sir Arvi, who worked alongside Sir Lindesay for many years, says: "He was a cultured person and a strong patriot dedicated to developing Australia; just making money would have been vulgar to him. He and the leader of the Country Party, Sir John McEwen, were friends and had very similar views."

Soon after control of the company was placed with Australian shareholders, Robinson made a number of requests for a special issue of 250,000 Western Mining shares to the Zinc Corporation, the centre of his Robinson's web of mining companies.

Sir Lindesay recalls in his memoirs, *Built on Gold*[5], that he refused these requests. "Had the Zinc Corporation retained WMC shares acquired at the conversion from GEFCA, the extra 250,000 may have enabled them to control WMC and thus transfer control of the Company back to England ... If these attempts to secure control had not been successfully resisted, much of the profits of WMC's subsequent successes could have been returned to England and lost to the Australian shareholders."

Clark had spurned Western Mining's founder, and now had the opportunity to build the company for Australian investors according to his vision. Clark and Robinson shared a passion for exploration using the best science, but they also had some important differences. Clark was more patient, and was happy to measure progress in decades, not years. Robinson seemed always to be searching for the next big win, the encore to the windfall made at Broken Hill all those decades earlier.

Clark also had a much wider vision for Western Mining. By the time he took over as chairman in 1952, he was already thinking about taking Western Mining into exploration for minerals other than gold.

The price of gold had been fixed for decades without change, but the costs of mining had marched relentlessly higher. In 1950, costs soared because of a worldwide rise in inflation that had its genesis in the Korean War. The US Government began stockpiling wool for uniforms and blankets, and drove up wool prices overnight by as much as 300%. The inflationary effects were felt across the world, including the Australian gold mining industry. Profit margins were squeezed further, and Western Mining's plans to diversify gained more urgency.

For Clark, diversification could only be achieved through exploration. New programs were begun in the 1950s to find alumina, copper, uranium, coal and oil and gas, despite

---

[5] Hill of Content Publishing Co, 1983, copyright G. Lindesay Clark

the shrinking financial resources that were available to fund these ambitious new searches.

Sir Arvi says Sir Lindesay set the tone of Western Mining for the best part of 40 years. "His great interest was not in production or management, but in minerals exploration and broad strategy. Whenever he came to visit the operations, he would spend all of his time with geologists, talking to them about their exploration plans and results. It was actually a problem for the mine managers and metallurgists. On one visit, I said to Sir Lindesay: 'Why don't you spend some time with the production people? They would like to show you what they do and tell you about their activities.' He would spend a day with them, but for the rest of the week he was back with the geologists. Then on the next visit he forgot about it altogether."

But Sir Arvi says Sir Lindesay was responsible for allocating the exploration funds that made Western Mining into a substantial, diversified minerals producer.

"It was largely due to him that funds were allocated for bauxite exploration in the Darling Range and nickel exploration at Kambalda at the time when even these very modest sums were a strain on the company's finances. He led the negotiations which resulted in the establishment of Alcoa of Australia and insisted that the initial nickel developments should be financed largely by equity raised by rights issues rather than loans, the wisdom of which was seen when the nickel markets turned down following the oil price increases in the early 1970s. The Company's growth from a small gold mining company into a substantial diversified minerals producer was largely due to his leadership. The discovery of the massive Olympic Dam copper-uranium orebody in the 1970s followed soon after his retirement."

⚒ 5 ⚒

Modern leaders

By the early 1960s, Clark had spent three decades making Western Mining in his own image. He was a mineral explorer by nature, and so had become the company. But the years were catching up with Clark, who reached the formal retirement age of 65 in January 1961. After a decade as both chairman and managing director of Western Mining, it was time to appoint new leaders who would carry on his unconventional vision for the company.

In 1962, he took his first step back from day-to-day control with the succession of W.M. (Bill) Morgan as managing director. Clark, who was known as Sir Lindesay after receiving a knighthood in 1968, would still come into the office every day, and would continue as chairman until 1974. He had known Morgan, a mining engineer, since the 1930s when Morgan worked for Gold Mines of Australia at Mt Coolon, a small, rich gold mine in Queensland that provided GMA's first source of income. Morgan subsequently transferred to Western Mining's operation at Bendigo, but left within a few years when the mine ran out of ore in 1937. He rejoined the company as a general manager in 1956, having distinguished himself elsewhere in the mining industry over the previous two decades.

Morgan proved to be an excellent managing director. He faithfully followed Sir Lindesay's blueprint for Western Mining as an exploration-led, diversified mining house. In his 10-year term at the top, Morgan successfully lobbied the Australian Government to lift its ban on iron ore exports and led the development of the Koolanooka iron ore mine. He also introduced Western Mining to the unfamiliar business of selling minerals on world markets. Until this time, all of Western Mining's revenues were from selling gold to the Perth Mint at a fixed price. Morgan had been a frequent visitor to Japan by the time he rejoined Western Mining and was ideally qualified to negotiate Australia's first iron ore sales to Japanese customers.

Bill Morgan's successor, Sir Arvi says: "Bill, like Sir Lindesay, was an urbane man with wide interests and a relaxed management style who disliked paper work and gave the people reporting to him wide autonomy as long as they produced results. Personal trust was the basis of relationships, as it had been in Western Mining since the beginning."

In the 1960s, Sir Lindesay was also grooming new leaders from inside the organisation, some of whom would have a huge influence on the discovery and development of Olympic Dam. These included Arvi Parbo, an Estonian-born mining engineer who joined Western Mining only a few years earlier in 1956 as underground surveyor at Bullfinch, a gold mine in the Yilgarn district between Perth and Kalgoorlie.

Parbo had just turned 30 and was making a late start to his career. He was a decade older than his peers when he finished studying at The University of Adelaide in 1955, but graduation as a mining engineer at any age was a remarkable achievement by the former war refugee. He had arrived in Australia in 1949 — alone, speaking minimal English, and not long recovered from a double hernia, a souvenir of heavy labour in a German coal mine.

Parbo was one of thousands of migrants arriving in Australia under government sponsorship. While many countries, including the US, England, Canada, Brazil and South Africa, had programs to resettle Europe's war refugees, Australia was the only nation not restricting numbers. "They would take anybody as soon as they could get them," he says. In return for the price of his passage to Australia by ship, Parbo was obliged to spend his first two years working wherever the Government directed.

He worked first as a jackhammer operator at a quarry in Marino, about 20 kilometres south of Adelaide. He had already taught himself to read English by avidly studying cheap paperbacks the US military issued soldiers in Germany after World War II. But the English he learned was nothing like the words bandied about by Australian quarry workers half a world away. Parbo's frustration had turned into a rare period of despondency, when suddenly his ear tuned into the rhythms and peculiarities of English as spoken by Australians.

By the middle of 1950, Parbo talked the Commonwealth Employment Service into finding him a job in a factory closer to the city so he could work full time and attend evening classes at The University of Adelaide. He was keen to resume studies in mining engineering that he began in Germany. Six years after arriving on a migrant ship in Melbourne, Parbo graduated in Australia in his chosen profession while still a relatively young man.

He had also married Saima Soots, another Estonian resettled in Adelaide as a Government-sponsored migrant. Soots and Parbo were among 80,000 Estonians who fled their country in September 1944 ahead of an advancing Red Army. Soots had fled with her mother, but Parbo was separated from his entire family in the scramble to leave Estonia.

Many Estonians headed for Germany. Even though it was still under bombardment by the Allies, Germany was a much better option than living under Stalin's murderous regime, which Estonians had already experienced in 1940. Within the first year of Soviet occupation, 60,000 Estonians — about 6% of the entire population — were forcibly removed. Among this number, almost 10,000 were executed, died in jail or disappeared without trace. Hitler drove the Soviets out of Estonia in 1942, but now the Red Army was returning.

Parbo and Soots had met briefly in Germany in a United Nations Relief and Rehabilitation transit camp. By chance, they would come together again at the Estonian Society in Adelaide, where they became folk dancing partners.

As Parbo's studies drew to an end, the head of mining engineering at the University of Adelaide, John Morgan, urged him to apply for a position at Western Mining because he believed the slow-moving and bureaucratic ways of an established mining house could snuff out his career before it even started.

"John took a big interest in his students, helping to bed them down into jobs in industry. He said to me: 'What you've got to realise is you are 10 years older than the other fellows going out. When you go into a large company, they will put you through a program for new graduates and by the time you get through that you will be 34 or 35. You can't afford to wait that long. You are better off going to small companies that don't have these kind of formal programs. And if you go to a company that is successful and develops quickly, they will want people for their expansion. You will be given a job with real responsibility whether you like it or not, and quickly. So don't think about Zinc Corporation or New Broken Hill, they are good companies, but not for you.'

"I said: 'That makes sense to me, but I don't know any companies. What do you suggest?' 'There's a company called Western Mining, it's pretty small, a gold miner in Western Australia, but I have been watching them. I like the people there, the fellow who runs it, Lindesay Clark, and Frank Espie, who runs the operations in Western Australia. And there's a fellow called Brodie-Hall who comes across as a pretty innovative and active fellow. This might be a good place for you to start.'"

Morgan organised a job interview for Parbo with Frank Espie, the deputy managing director, who regularly stopped in Adelaide on his way between Kalgoorlie and board meetings in Melbourne. On the appointed day, Espie was running late for a more important meeting, so the job interview was held on the corner of King William Street and North Terrace as Espie walked towards his next appointment.

"He said 'I understand you want to work for Western Mining?' I said: 'Yes, Mr Espie.' He said: 'Done!' John had given some information about me; it wasn't completely uninformed, but it was no interview."

Parbo's career with Western Mining progressed quickly, just as Morgan predicted. After two years as underground surveyor at Bullfinch, Parbo was promoted to underground manager at the nearby Nevoria mine. After another two years, Parbo was transferred to head office in Melbourne to take up the role of technical assistant to the general manager, who was then Bill Morgan. Parbo would also frequently be assigned to tasks for Sir Lindesay.

The role was supposed to last one year, but Parbo remained for four years, providing much-needed assistance at a time when Western Mining's joint venture with Alcoa was getting off the ground. It was to be a very formative time in Parbo's career.

Sir Lindesay had created the role in 1949 to provide him and senior managers in head office with technical support. Head office consisted of only about 20 people in administrative functions. The company's technical and operational centre was in Kalgoorlie. By rotating up-and-comers in this role, Sir Lindesay and senior management could also get a close look at potential future leaders of the company. At the same time, youngsters like Parbo would gain first-hand instruction in Sir Lindesay's distinctive values and ideas, and hence those of the company.

In the parlance of modern management, Sir Lindsey would probably be said to be sharing organisational culture, managing talent and succession planning. But there was nothing so fancy in the 1960s, according to Sir Arvi. "There was never any formal discussion about values or culture. The commitment to exploration was an unquestioned part of Western Mining. We did not have to talk about such things. The company was still small enough for everybody in management to know each other and understand how things were done. We were short on processes and procedures and long on getting things done with the minimum of fuss. We did know that this was different from how many other companies operated and regarded it as an advantage."

He says his time in head office confirmed what he had already come to know about the company: exploration was the company's major focus. "Today, many companies think their business is making money. We thought our business was finding ore bodies and mining them. If we did that well, then we would make money."

An amusing anecdote from the early 1960s, taken from Sir Arvi's 2007 oral history with the National Library of Australia[6], reveals much about his character and at least one of the factors in his smooth transition to life in Australia. Not only had he kept a sense of

---

[6] National Library of Australia Oral History, Sir Arvi Parbo interviewed by Rob Linn, 15 February 2007, TRC 5771

humour, despite all the traumas of Europe, it was a style of humor that quickly reassured Australians this strangely named foreigner was just the same as them.

The story is from April, 1961, when Sir Lindesay set off for Pittsburgh to talk to the mighty Alcoa company about developing the bauxite deposits recently discovered by Western Mining near Perth. Parbo was asked to accompany the chairman and his wife because of his knowledge of the discovery and the proposed project. Sir Arvi begins the story by saying that in the early 1960s there were no fax machines, let alone emails.

"So telegrams were the normal thing, and when the arrangements had been made a telegram was sent to say, 'MR AND MRS CLARK' ... he was then Mr Clark ... 'AND A PARBO arriving Pittsburgh' with all the details. And later we found that ... quite a bit later actually ... they didn't tell us until we got to know each other pretty well, this caused great confusion in Pittsburgh because the telegram was all capital letters and the full stop had been left out, so the message read, 'MR AND MRS CLARK AND A PARBO ARRIVING IN PITTSBURGH,' and they didn't know what 'a parbo' was. Nobody knew what 'a parbo' was. [Laughs]

"So, apparently they had several executive meetings trying to work it out. They didn't know whether it was Mrs Clark's pet poodle or whatever. Anyway, they finally came to the conclusion that whatever 'a parbo' was, it should be booked a room at the Pittsburgh Hilton, and they did."

After four years in Melbourne, Parbo returned to Western Australia in 1964 to take up the role of Deputy General Superintendent, based in Kalgoorlie. He was now second-in-charge of all operations behind Laurence Brodie-Hall (later Sir Laurence), who had recently relocated to Perth because he was spending much time dealing with the West Australian Government on various projects and operations.

The discovery of high-grade nickel mineralisation at Kambalda in 1966 was to be a turning point for Western Mining and for Parbo's career. The company had decided to rush its Kambalda nickel project into production, fearing the nickel boom could bust at any time. If it took too long to develop Kambalda, the company would lose an opportunity to generate substantial profits for the first time in its history. It was also a rare opportunity to break into the world nickel market, which until that time was dominated by a duopoly of Canadian companies. Parbo was a key member of the Western Mining team that built the mine, the processing plant and a township for its workers in only 17 months, establishing a position in the nickel market and creating a profit boom for Western Mining before the market eventually turned down in the mid-1970s. By 1972, Western Mining had also built a nickel smelter in Kalgoorlie and a refinery in Kwinana, south of Perth.

Elsewhere in his oral history, Parbo says: "... the Western Mining board, recognising the realities, decided that speed of getting into production was of absolute superior importance to anything else, nothing else mattered. So we were told, and I was very much involved in it at the time, getting this project going, to forget about plans or budgets of any of those things. Just start producing nickel. And we did that....we did have a plan, of course. We drew one up quickly. We had to decide where to build the plant ... and where to sink the shaft and drill more holes. But these were the sort of things which kept changing all the time. Whatever the plan was last night might be quite different to the plan tomorrow or the day after. It kept changing as the information kept coming in from exploration. And the ore bodies kept getting bigger ... and we would amend our plans in a sort of continual manner. We didn't have staff meetings. Staff meetings would be held standing up in the corridor. We didn't have time to sit down literally. There were no weekends and no holidays. None of those things. And the upshot of it was that we produced the first nickel concentrate 17 months after the first drill hole intersection."

Parbo was on a midnight plane to Melbourne on the day the plant officially opened on 15 September 1967. He was appointed general manager, second in charge to Bill Morgan, a few months later in February 1968. He would soon be thrust into the role of managing director when Morgan's life was cut short by cancer in 1971. Parbo was only 45 years of age. He was an unusually young CEO by the standards of the day, and a foreigner at that. But Western Mining had never had any prejudices about age, nationality, beliefs or even gender (one of its first geologists in the 1930s was Florence Armstrong).

Parbo's rise to the top of Western Mining attracted a great deal of media interest; he was perhaps the first of Australia's great migrant success stories to emerge in the 1970s and his success made everyone proud. Media attention was then ongoing because of his outspokenness on the mining industry and national economic issues. If he saw something out of step with reality and counter to the creation of jobs and wealth, Parbo would say as much publicly. He was unusually frank for an Australian captain of industry, but Parbo saw it as his duty to speak up on behalf of common sense.

Parbo handled the media limelight with ease because his views and his words were always well considered, and always his own. In the 1970s Western Mining did not have economists or government relations advisers telling the chief executive what to think or say. The company did not even have a public relations department until some time in the 1980s. Parbo, who was knighted in 1978, wrote many of his own speeches. His written English has beautiful economy and precision, which you might expect of an engineer, but not one who counts English as his second language.

Sir Arvi modestly claims today to be a minor player in the Olympic Dam story, and that he simply followed the principles set out by Sir Lindesay for running Western Mining. In fact, he had a profound influence in the years leading up to the Olympic Dam discovery in 1975. His contribution began as early as 1968 when, as general manager, he was influential in granting approval for an autonomous exploration division.

Earlier attempts by the company's exploration leaders to gain control of their own budgets and exploration priorities were declined. But in November 1968, Parbo wrote to the company's ambitious new chief geologist, Roy Woodall, granting permission to implement his plans for a new Exploration Division. The division's budget would be approved in terms of expenditure for each of a number of identified main projects, with additional amounts for reconnaissance as well as divisional management. Within the total approved budget, Woodall had freedom to move funds from one project to another if exploration results warranted. Management and control of spending were now the responsibility of the Division.

There were a number of factors in the company's change of heart on this issue, including the end of a long-standing rift between the former chief geologist, Don Campbell, and the company's most senior executive in Western Australia, Brodie-Hall. The departure of Campbell removed one of the barriers to an autonomous exploration division, while Parbo's recent rise into senior management was another positive factor. As Brodie-Hall's deputy in Western Australia between 1964 and 1967, Parbo had first-hand knowledge of the practical difficulties created by the old exploration division structure.

Sir Arvi Parbo in 1979

Sir Arvi therefore was much more than a follower of Sir Lindesay's principles; he was instrumental in giving Exploration Division more freedom than it had ever enjoyed. Now formally invested with the trust of the company's directors and management, "Ex Div" became an environment in which the boldest ideas of the company's top geologists could flourish and be tested. These included an ambitious, world-first model for copper discovery that would uncover Olympic Dam.

Sir Arvi's support for the exploration effort was just the beginning of his contribution to Olympic Dam. He would play a leading role in the 13 years between the first discovery hole and the commissioning of a mining, processing and refining operation.

There were perhaps a dozen major barriers to the development of Olympic Dam, not least of which was a South Australian Government against the mine. The company faced major technical challenges in developing a complex ore body that had no precedent anywhere in the world. Then there were the financial demands — greater than most resource companies in the world could handle, and certainly beyond the ability of Western Mining to proceed on its own.

Against these odds, many companies might have made a quiet exit in favour of a much bigger, new owner that was better equipped to deal with all the problems. But Western Mining had taken 20 years to find a copper deposit, and Olympic Dam was its most brilliant exploration success. The company would give away gold bars in the main street of Kalgoorlie before giving up on Olympic Dam or signing it over to another company to develop.

Western Mining also had the advantage of having Parbo at the helm. No-one could be better equipped at dealing with tough realities and overcoming adversity through thoughtful and persistent effort. He also had time on his side. When Olympic Dam was discovered in 1975, Parbo was just four years into his term as managing director; he was merely getting started at a company where the average term of the chief executive was measured in decades, not years.

Roy Woodall

Western Mining in the 1960s also produced a great leader of exploration, Roy Woodall. If W.S. Robinson could have conjured his ideal geologist, he would most likely be Woodall. In fact, the two men did meet in the 1950s, and Woodall often quotes Robinson's edict of using the best science to find what might be.

It would certainly be hard to meet someone more passionate than Woodall about science and its potential to unlock mineral wealth. His intensity on the subject is matched only by his blue eyes. Meeting him makes you wonder whether he could actually see into the Earth and find orebodies without the assistance of any remote-sensing technologies. Of course, there were no such short cuts to Woodall's incredible record of success as an exploration geologist and leader of Western Mining's exploration division between 1967 and 1995. He had to learn the science the same way as everyone else, but a few things set Woodall apart. From the beginning of his career, he had a hunger to learn the very best science in the world and to bring this knowledge back to Australia to make discoveries. Woodall was also a gifted at recruiting other talented people and marshalling their skills, ideas and enthusiasm. This was to be his most important contribution to the discovery of Olympic Dam.

Woodall's oral history, recorded in 2004 by the University of California, Berkeley[7], reveals he was born with a love of science, but he did not always feel geology was his particular destiny. As an undergraduate at The University of Western Australia in the 1950s, he took on additional studies in chemistry and geology so he could defer a decision on his major discipline until his final year.

"... It was a difficult choice, but what probably persuaded me to choose geology were two things. The first was the opportunity not to be laboratory bound. You could do laboratory work in geology, but you needed also to get out into the field and observe the rocks in their natural setting. And the second reason was my fascination with the mining industry, which was so linked in importance to the whole history of the economic development of Australia, much as it was to the development of the United States. Why was it important? Because it generated so much true wealth: wealth that the nation didn't know existed until the ore deposits were discovered. Wealth that saved many of the state governments of Australia from going bankrupt in the early colonial days. Wealth that saved the country in the Depression years

---

[7] BANC MSS 2008/205, Australian geologist, 1953 to 1995: oral history transcript: success in exploration for gold, nickel, copper, uranium and petroleum / Roy Woodall, The Bancroft Library, University of California, Berkeley

when gold mining became a very important form of employment and support for families. Wealth that had made a tremendous impact on the whole economy of Australia, and still does."

Growing up in Perth in the 1930s, Woodall had personal experience of the hardship created by the Great Depression. He was forced to leave high school after only three years to find work, barely 16 years of age. He finished his secondary schooling by attending evening classes over two years.

In 1952, Woodall completed an Honours year of his Bachelor of Science Degree. He so impressed the earth science faculty at The University of Western Australia that it offered the opportunity to begin a PhD the following year. But the 22-year-old first wanted a taste of the industry and began work at Western Mining in April 1953. As an undergraduate, he had worked for the company over summer holidays, and was impressed by the company's emphasis on top-class science.

A year later, Woodall was hungry for more knowledge. Determined to be exposed to the very best geoscience in the world at that time, he wrote to half a dozen of the top universities in the United States, enquiring about scholarships that might pay his way while he undertook a Masters degree or PhD.

By mid-1955, he was sailing for San Francisco via England to undertake studies at the University of California, Berkeley. Most of his living expenses were covered by the English Speaking Union, which each year granted one scholarship to a graduate from New Zealand or Australia to study at Berkeley. His travel costs were covered by the Fulbright Organization, which even kicked in extra funds to meet Woodall's desire to travel via Canada, one of the world's richest mining countries. Many of Canada's famous

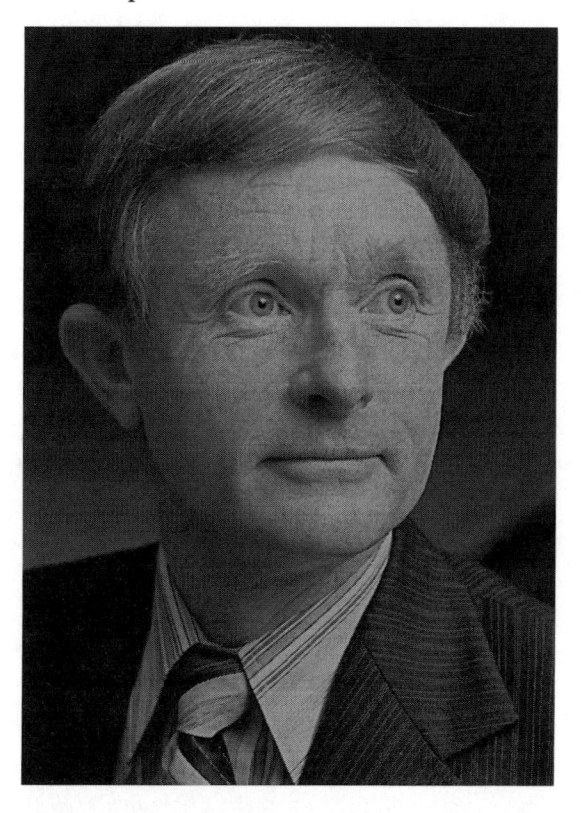

Roy Woodall in 1978

deposits of copper, copper-zinc and nickel were in a Pre-Cambrian-aged craton, not unlike the Yilgarn Craton that hosted the gold deposits of Kalgoorlie.

Western Mining did not make a financial contribution to Woodall's overseas study, but Frank Espie, the man who would interview Parbo on an Adelaide street corner a year later, did help in an unexpected way. Not long before he was due to leave for the US, Woodall and a young woman from Kalgoorlie, Barbara Smith, became engaged.

"We ran into some problems because Barbara was only nineteen, and then the legal marriage age without parental consent was twenty-one. The only way I eventually persuaded Barbara's parents to approve our marriage was as a result of the intervention by ... Frank Espie. He was a very famous man. He had worked in Burma, in charge of mines there up until the Second World War. When the Japanese invaded Burma, he successfully evacuated all women, children, and men by walking them out carefully through the jungle through to India. It was a dramatic and famous trek. And he was that sort of a leader. He succeeded in persuading Barbara's parents and we were married on May 21, 1955."

Woodall sailed for England within the fortnight, with his new wife to join him in San Francisco as soon as he was settled. Woodall says the decision to study at Berkeley was one of the most important decisions he made in his life. Later, he would make it possible for many of his most talented geoscientists to repeat his experience by encouraging them to study at the world's leading universities. He introduced a study leave program that would allow staff to spend years away from Western Mining, with their position held open until they could return. The scheme even allowed for many of those on study leave to receive half of their Western Mining salary while undertaking postgraduate studies.

Woodall returned to Western Mining in 1957 with his young family (he now had two children, the first of 10). He was promoted to assistant to the chief geologist, Don Campbell, and was over-flowing with excitement about new ideas for mineral exploration in Australia.

"I was absolutely convinced, especially after having seen the great ore bodies in the Pre-Cambrian of Canada, that there had to be ore bodies of a similar type in the rocks in Western Australia; there had to be ore deposits there other than gold deposits... They couldn't just be gold-bearing when the same rocks in Canada contained deposits of gold, nickel and copper, zinc. At least that's what Berkeley taught me!"

Sir Lindesay and Don Campbell were supportive of Woodall's new thinking. Profits from gold mining had dropped sharply in the 1950s, and Sir Lindesay had already determined to diversify Western Mining into other minerals. However, the company's small financial resources meant non-gold exploration would be very modest. There were no funds to establish a world-class research or mineral intelligence department, as Woodall had proposed in memos to his bosses even before he returned from Berkeley.

Woodall's resumed his career at Western Mining in spectacular fashion, although he says it did little to help his career at that stage. "Mr. Clark reminded Mr. Campbell that bauxitic laterites were known near Perth in Western Australia. Don Campbell showed me the letter and said, "What do you think, Roy? What do you know about these bauxitic laterite deposits near Perth? After all, you studied at the University of Western Australia nearby." I knew a little about them, but I did what I thought was the only smart thing to do. I looked for information in the literature, and I came across a report: Report (No. 24) by the Bureau of Mineral Resources, a federal organization like your United States Geological Survey. They had examined those deposits, confirmed that they were bauxitic, but uneconomic because they were low-grade and had a high content of silica."

The high silica content was considered to be the biggest problem with the Darling Range deposits because clay-type silica consumes large volumes of caustic soda, an expensive but essential part of the process to upgrade bauxite to alumina. "But something was being overlooked! What was important about the bauxitic laterite near Perth was how the silica occurred. I found this Bureau of Mineral Resources report, and it reported the silica in the laterites near Perth occurred almost entirely as quartz, not in clays. Well, I said to Mr. Campbell, "There is no reason why these deposits shouldn't be excellent bauxite because the quartz is just a diluent and it should not dissolve in the refining process, it should not consume caustic soda, and if removed, will enhance the grade."

The Darling Range bauxite deposits would be developed by Western Mining in conjunction with Alcoa of America, and today still provide more than 10% of the world's alumina supply.

Woodall says he was disappointed not be involved in the evaluation of the deposit, and still didn't get permission to set up his proposed department to look for other minerals in Australia. Instead, he was sent to Tarraji River in the remote Kimberley region of Western Australia to search for copper. The project was unsuccessful, but it marked the beginning of long search for copper that would lead 20 years later to Olympic Dam.

Woodall spent time on various gold projects over the next few years and even on a talc deposit that would become a small but very profitable mine for the company. But almost 10 years after returning from Berkeley with a burning desire to search for base metals, there was still no money for such a project.

Woodall then changed his own fortunes and the future of Western Mining by devising a low-budget plan to investigate what he believed to be a weathered outcrop of a nickel orebody at Red Hill, about 50 kilometres south of Kalgoorlie. The location would later become known as Kambalda.

Assays of outcrop samples brought to Woodall by prospectors revealed significant percentages of nickel, along with 100-times the expected level of tellurium. These results immediately excited Woodall because he knew from his studies at Berkeley that tellurium was a typical indicator of nickel-copper sulphide deposits in Canada. He was convinced that a high-grade nickel ore body would be found below the mineral-leached outcrop.

Woodall visited the outcrop and was able to trace it over a distance of 1,200 feet. This additional news finally generated some financial backing, although it was still only about $20,000 in today's terms — enough to hire two undergraduates over the summer holidays to map the full extent of the outcrop and assay some samples.

The mapping revealed the potential for nickel ore bodies over a distance of 20 kilometres. Funding for a six-hole drilling program followed quickly, and the first hole in January 1966 returned a spectacularly rich intersection of 8.3% nickel over a distance of more than two metres. News of the first nickel discovery in Western Australia created a frenzy among explorers and sharemarket investors. This became famously known as the Poseidon boom after an Adelaide-based mineral explorer. Poseidon shares soared from a few cents to a peak of $280 (in 1969 dollars) on the strength of a nickel discovery at Windarra, several hundred kilometres north of Red Hill (Kambalda), based on a prospector's find of the same type of leached outcrop.

Western Mining's own share price increased 10-fold because of the ultimate success of Woodall's dogged pursuit of base metals around Kalgoorlie. He would not have trouble getting funding for any project for the rest of his career. The Red Hill (Kambalda) nickel discovery also won Woodall the respect of geoscientists around the world. While his North American experiences provided invaluable clues to the presence of high-grade nickel minerals, Woodall had actually uncovered an entirely new type of nickel ore body, and the first in Archean-aged rocks anywhere in the world. The deposit was associated with a strange type of rock that became known as komatiite. The rocks are believed to have formed when extremely hot lava, at temperatures of 1400 to 1500 degrees Celsius, erupted below a shallow sea. The immediate quenching of the lava gave the rocks a spinifex-like texture.

With the discovery of three more nickel deposits at Kambalda in 1967, a bright future for Woodall was assured. At the beginning of 1967, he asked Western Mining for a year of study leave to revisit academia and update his knowledge of the best science now being conducted in his field. Managing director, Bill Morgan, granted him a year's leave on full pay, and the company even paid the return travel costs to North America for Woodall and his entire family, which had now expanded to eight children.

Don Campbell resigned in mid-1967, and when Woodall returned to Australia at the end of the year he was the obvious candidate to take charge as chief geologist. He inherited what was already the most professional exploration outfit in the country, a legacy of the Harvard geologists who set the foundations in the 1930s. The maps produced by Western Mining geologists were famous for their meticulous detail, with greater use of symbols to record more information, more accurately than was the standard of the day. Woodall had learned his craft in this environment, and now as chief geologist constantly told his staff to record only the facts of what they see, not their interpretation. Assumptions could be false and bury a crucial fact that might be the difference between discovery and despair.

In 1968, Woodall's Exploration Division was given complete autonomy to operate within the budget set by the company's directors. The company's finances still had to be carefully rationed because of large bank loans on the second stage of the Kambalda development, but everything in the company was suddenly much bigger because of the nickel success.

The 1968 exploration budget was increased three-fold from the previous year to $15 million in today's terms. A large component of this budget would be spent on recently hired geologists who were searching for additional nickel and gold ore at the company's existing mines. The number of geoscientists on staff had jumped from only a dozen in 1965 to 20 in 1966 as a result of a recruiting drive that began with the first nickel discovery in January of 1966.

However, there would still be room in Woodall's expanded budget to start spending money in the ways he had always dreamed since returning from Berkeley in 1957. One of his first moves was to establish regional exploration offices in Western Australia. Until now, all exploration was based in Kalgoorlie. The new offices, which comprised one or two geologists and a field assistant, were at Leonora, Meekatharra, Wiluna and Wittenoom, in the heart of the Pilbara region and some 1,400 kilometres north of Kalgoorlie.

"Now they were being domiciled, resident, 300, 400 or 800 miles away, in small, isolated towns. The philosophy or strategy was; let's get our good geologists close to where the mineralized districts are. There they are much more likely to become very familiar with the geology, familiar with any prospectors that might be in the area, and feel that this was their province, where they were fully responsible. It was a matter of giving them responsibility, and motivating them to be good explorers."

With further increases in exploration budgets, Woodall would extend the idea of regional bases to other states of Australia. Exploration activity outside of Western

Australia would be led by an Eastern States Exploration Division, based in Melbourne under Jim Lalor. As the immediate leader of the search for Olympic Dam, he would have a crucial role in its discovery.

The expansion of geographic coverage under Woodall was not a pivotal development; the company had explored outside Western Australia for many years. But Woodall did make big changes in the way exploration activity and the company's geoscientists were managed.

Woodall says in an interview for this book that recruitment is the starting point of exploration success. "I spent one third of all my time on recruitment and perceived it to be my most important function. I only wanted good scientists with a great drive to go and find something."

"If they asked me in interviews whether the company had a bonus system, I automatically ruled them out of a job."

Having geologists he could trust to a make great exploration effort was critically important. "Managing exploration is not like running a factory. Your people are out in the boondocks and you can't keep an eye on them. You have to be able to trust your people."

Once you became part of Western Mining Exploration Division, you belonged to what Woodall liked to refer to as a guild. His geoscientists were like craftsmen in an association, run on the principles of mutual aid and the pursuit of a common goal. Many of the people who worked in Exploration Division under Woodall say this was indeed how he treated his people and managed the entire team.

Dan Evans, who would establish an Adelaide office for the Exploration Division and become a key member of the team that found Olympic Dam, says Woodall established a system of six-monthly technical reviews that formalised the frequent sharing of ideas and expertise on every project.

"It was really important in the search for Olympic Dam. The six-monthly reviews would feed in the ideas of our internal experts in every discipline, including field geology, ore deposit modelling, geochemistry, geophysics and tectonics. The reviews encouraged a lot of challenging discussion and thinking, which ultimately helped us make what we believed were the best decisions about exploration projects."

Woodall took a close interest in the career of every geoscientist. Once a year, he would sit down with each individual or, if time or distance did not permit, with each individual's direct supervisor. These annual career reviews would check the individual's progress over the past year, and determine how his or her responsibilities should be changed or expanded in the next 12 months to develop their career.

Scientific staff were encouraged to attend conferences, and in-house courses were frequently arranged. One or two lucky individuals each year were offered the opportunity of study leave of a minimum of one year at half their current pay. Woodall took the lead in recommending where the individual should study. Typically, he would be pushing for a leading overseas university, where his scientists would be exposed to the latest and best science. Study leave was an exceptionally good opportunity because Woodall's international reputation opened doors to top universities that would otherwise be closed.

Woodall also encouraged camaraderie among his team. In 1966, while still assistant to the chief geologist, he organised the first annual dinner in Kalgoorlie for Western Mining geoscientists, Kalgoorlie administrative staff and their partners. The dinner was part of a weekend of functions, some of which were serious but most were simply an opportunity to let off steam after a year of hard work.

In the early years, those who were not familiar with the annual dinner could receive a rude shock at the door; Woodall would wait with a pair of scissors and deftly remove the neckties of guests who made the mistake of wearing anything so formal. It was a larrikin act that got Woodall into hot water on at least one occasion, but it did make a powerful statement about the culture of Western Mining. To be part of this company, you had to thumb your nose at formality, at convention and at the established order. And that's exactly how Western Mining had carved out a place for itself as major player in the Australian resources scene by the late 1960s.

Exploration Division cemented its reputation as the country's greatest exploration outfit with another major discovery in 1972, the Yeelirrie uranium deposit, about 500 kilometres north of Kalgoorlie. The discovery was based on new thinking by Dr Eric Cameron, an Englishman recruited by Woodall to find a uranium deposit. Yeelirrie was the first uranium discovery of its kind in the world and sparked a rush of uranium exploration activity in the region.

Yeelirrie is a deposit of world scale and Western Mining invested many millions getting to a position where a development decision could be made. The Hawke Labor Government finally scuttled the project in the 1980s, but its attractive economics and size were recently confirmed by BHP Billiton, which is pushing ahead with a new development plan.

The company's emphasis on innovative exploration was paying huge dividends. At the start of the 1970s, in the years immediately prior to the search for Olympic Dam, the board's confidence in its geoscientists had never been higher.

6

REINVENTING THE SEARCH FOR COPPER

By the early 1970s, Western Mining's exploration division was a formidable machine. The company had found a wide range of economic mineral deposits since making the decision in 1953 to diversify from gold. With discoveries of bauxite, iron ore, nickel and uranium, it seemed to the rest of the world that Western Mining could summon orebodies almost at will.

But one commodity — copper — still eluded its geologists, despite intensive exploration for 15 years and the expenditure of many tens of millions of dollars in today's terms. Like many other mineral explorers in the 1950s, Western Mining was keen on copper. The economic boom that followed World War II created a surge in demand for electricity and the copper cables that could carry it into homes and factories. Developing countries such as Japan, South Korea and Taiwan also became huge consumers of copper for the electronic devices they exported to the increasingly comfortable middle classes of the Western world.

Western Mining's copper search began in 1957 in the remote Tarraji River area, about 350 kilometres up the coast from Broome in Western Australia's Kimberley district. Some shows of copper found by a prospector, Bill Rogers, attracted the company to the area. Tarraji River was Roy Woodall's first major assignment after completing his Masters degree in the US, and a world away from the famous wooded campus of the University of California, Berkeley, on San Francisco Bay.

Woodall says in his oral history that Tarraji River was the most remote and difficult location in which he worked. "It took a whole day to go from the camp to Derby, the nearest town, to get provisions, which we did every second week. All meat was obtained by shooting the wild cattle that were roaming the country. We almost lived off the land, in a way."

The only communication with Kalgoorlie was by radio through the Royal Flying Doctor network.

The three-year exploration program at Tarraji River found some copper, but the deposits were too small to be economic, especially considering the economic hurdles created by the remote location.

Woodall says he applied the best science and some innovative techniques, including a new field test for geochemical sampling. "With the Bureau of Mineral Resources, we developed a colorimetric test for copper, lead and zinc to use in the field. It was a crude but effective way to identify anomalous amounts of copper in the field, with the hope of tracking them upstream to their source."

Tarraji River was also his first opportunity to apply geophysical survey methods. "It so happened that in 1957, a Yugoslav immigrant arrived in Kalgoorlie looking for work. He was obviously very intelligent. He said he was a geophysicist, but we had no way of knowing how much geophysics he knew because he could hardly speak English. His name was Anton Triglavcanin. Initially, we just employed him as a draftsman until he could speak English well enough to tell us what he really knew, and he turned out to be a brilliant geophysicist. And so Anton and his wife, Mary, came with me in 1958 and lived in a tent camp in the wilderness where he experimented with some of his geophysical techniques."

As the search at Tarraji River drew to a close, Western Mining entered another copper exploration program on the other side of the Australian continent. The new program centred on South Australia's historic copper mines of Moonta, Wallaroo and Kadina, at the top of Yorke Peninsula and about 150 kilometres north of Adelaide. Western Mining began exploring in the region in 1960 after acquiring exploration licences over a vast area of more than 40,000 square kilometres. The project began as a joint venture with North Broken Hill and Broken Hill South, which each held a 25% interest.

The partners retained a Canadian company, McPhar Geophysics, to conduct a survey using its induced polarisation (IP) technology. The newly developed survey technique could reveal the presence of electrically "polarisable" copper sulphides by measuring properties related to rock conductivity and resistance of rocks at depths of up to 100 metres or more.

The IP survey revealed an extensive anomaly, and drilling quickly got underway in 1960. The first two holes produced "tremendously encouraging" intersections of high-grade copper, according to Douglas Haynes, who spent 1968 re-logging the many exploration holes that were drilled. But the complex copper-gold-magnetite mineralisation of the area defeated all attempts to find a significant economic copper deposit. By 1971, the joint venture had spent $21 million in today's terms without success. North Broken Hill took over as operator and began to fund all exploration costs, thereby increasing its equity interest in the project.

Western Mining's geologists would come across similar unusual geology many years later at Olympic Dam. The Moonta-Wallaroo mines were on the same eastern edge of the Gawler Craton as the Olympic Dam deposit, and contained a similar but distinctive iron-oxide style of mineralisation.

The Moonta mines even contained a significant amount of uranium, although the Cornish miners who came to the South Australian colony in the mid-1800s would not have known of its presence or its radioactive nature. The uranium produced by these mines still sits today in the waste rocks just outside Moonta, which show anomalously high readings in modern radiometric surveys.

The mines of Moonta-Wallaroo and Burra, about 160 kilometres to the east, produced 10% of the world's copper in the mid-1800s and saved the early South Australian colony from bankruptcy. Some 150 years later, the geologically related orebody at Olympic Dam has the potential for a modern-day transformation of the state's economy.

In 1965, Western Mining embarked on another new copper search after Woodall was shown rocks dug up by Pitjantjatjara Aborigines on a Native Reserve in the Warburton Ranges, a remote region in Western Australia, near the geographic centre of Australia. Some of these rocks were fantastically rich, assaying 60% copper as well as having high grades of silver. But what really attracted Woodall's attention was the style of copper-hematite mineralisation in these very old, altered basalt rocks. He noticed the similarities with the famous deposits of the Keweenaw Peninsula in Michigan, and decided to take this new find seriously.

Woodall visited the Warburton Ranges in July 1965, and suggested to the local superintendent of native welfare that Western Mining train and support the local Aborigines in their small-scale copper mining venture, which had occurred spasmodically for a number of years. In return, Western Mining wanted permission to explore the area, as he recounts in his oral history: "We decided that we would send to this remote Aboriginal community, five hundred miles from Kalgoorlie, out in the desert, a complete mining party of four people: a registered shift boss so mining would be legal and able to conform to the Mining Act regulations, and three good miners. We mined the rich veins the Aborigines had found, and we cut a three-way deal with the Aboriginals. We said, 'A third of the revenue is yours, a third of the revenue is for the mining party, and a third of the revenue is for Western Mining to compensate for the cost of the equipment and supplies which we supply.' Now we mined there for four years and prospected the area. Eventually we decided that even though the deposits were very rich, they were too small to be profitable in that very remote locality. We carried out extensive exploration in the district, looking for larger deposits, but failed. It was however a very well managed program: our geologist-in-charge lived in a comfortable mud-brick house which the Aborigines helped us to build, and we had a well equipped mobile analytical laboratory on site. We employed Aborigines and trained one to be a laboratory assistant."

In 1967, the company hedged its bets on Warburton by starting another major copper exploration project, focused on the Hamersley Ranges, about 1,100 kilometres north of Perth. This was another idea of Woodall's after a trip that year to see the great mineral deposits of Africa, including the stratiform copper deposits of Rhodesia. He knew Australia had sedimentary shales of the same type and geological age, so he reasoned that exploration might find great copper deposits of this style in Australia. The same type of thinking had led him to the nickel deposits of Kambalda, which first stirred in Woodall's mind after he toured the famous Sudbury nickel mines in Canada.

Soon after Woodall's return from Africa, a prospector showed him a shale rock with copper from near the town of Wittenoom in the Hamersley Ranges. A major new exploration base was immediately established in the town and would become known as the Fortescue Copper Project. It included an assay laboratory and a large team of geologists that could cover the vast tracts of land being explored.

Home-grown best science

By 1969, Western Mining had run hard without success for a number of years at three major copper exploration projects — Moonta, Warburton and Fortescue.

Total spending on the search for copper, including Tarraji River, had climbed to $21 million in today's terms, a huge expense for a company that had only begun to earn significant profits in 1968 with the start of its nickel operations.

Woodall had seized every opportunity to use the best science and innovative thinking, including new geochemical and geophysical survey techniques, and aggressive exploration for possible Australian analogies of copper deposits in the US and Africa. Now came an opportunity to develop some world-best science of its own. One of Western Mining's youngest and brightest geologists, Douglas Haynes, decided he would go back to university for three years to undertake a PhD in a certain style of copper mineralisation. Haynes had joined Western Mining in late 1966 straight from The University of Western Australia. He was like many university students of the Vietnam era, with long hair and outrageous sideburns, but Haynes was unusually bright — a gifted English student as well as a brilliant young scientist in the making.

Douglas Haynes pictured in 1970 while at work on his PhD.

The established and conservative mining houses of the 1960s would probably not have given him a job because of his appearance, but Western Mining had no such hang ups. If you were a smart geologist and excited by the idea of finding orebodies, you were welcomed into the Western Mining Exploration Division family.

Haynes joined in the first year of a massive recruiting drive by Western Mining. The discovery of nickel at Kambalda at the start of 1966 transformed the company and it suddenly needed every top quality geologist it could find. Haynes was one of four geology undergraduates completing an Honours year at The University of Western Australia, along with Geoff Hudson, Gordon Dunbar and Ian Campbell. Don Campbell, Western Mining's exploration chief and Ian's father, visited the Honours group in their UWA laboratory to interview Haynes, Hudson and Dunbar about joining Western Mining. He offered all three young men a job and they effectively signed up on the spot. They were excited about Western Mining's world-first discovery at Kambalda and eager to join such a progressive company. Ian Campbell was not present the day his father visited, but he also was recruited at about the same time to Western Mining's rapidly expanding Exploration Division.

Haynes began work in the first weeks of 1967, based in the headquarters of Exploration Division in Kalgoorlie. He spent three months in the nearby gold fields of Coolgardie studying the mineralogy of pegmatites, an unusually coarse-grained igneous intrusive rock with a similar composition to granite.

In March 1967, he was sent to join the copper exploration project in the Warburton Ranges, followed by six months on the Fortescue project between June and December 1967. The next year was spent at Moonta, trying to make sense of the many drill cores that had produced such disappointing and confusing results since the first exploration holes in 1960.

Haynes had always planned to go back to university for a higher degree, and after two years of extensive work on Western Mining's copper projects, the time seemed right. "I always wanted to work a few years just to get some practical rounding. Then I wanted to do a PhD or post-doctoral degree that was relevant to Australia's future. I didn't want to study esoteric, totally academic topics — I wanted to look at something that would contribute to mineral discovery in Australia.

"The thing that Australia was short of at that time was copper. As I had spent a lot of time working on copper mineralisation, it seemed like a good topic to study. Western Mining was producing nickel, and a lot of its customers wanted copper as well.

"I wanted to look at the genesis of copper deposits in the Warburton Ranges because I had done mapping in that area. The idea was to continue mapping in the area of mineralisation, collect samples of the altered basalt rocks and other rocks in the region and perform a geochemical study to look at losses and gains of different major and trace elements.

"The study would relate those losses and gains to the types of alteration minerals and then use those alteration minerals to set constraints on the hydrothermal ore-forming system. From there, the study would rig a generalised model of copper occurrence that could be applied in exploration in the future. That was basically the aim at the very beginning."

It was an ambitious study and certainly beyond the capability of most geology graduates, even with a First Class Honours year behind them. However, Haynes had the background to do it; he loved chemistry almost as much as geology and was still taking chemistry subjects in his final years as an undergraduate. He says geology won narrowly over chemistry as a career path because of his desire to be outdoors and the exciting prospect of being involved in mineral discoveries. It was a time when major discoveries of iron ore in the Hamersley Ranges and uranium in the Northern Territory had created great excitement about a resources boom.

The PhD study focused on the chemical reactions when magnetite in basalt altered to hematite. Both are forms of iron ore, but have very different physical properties. Magnetite is magnetic, hematite is not; magnetite can hold copper in its crystal structure, but hematite cannot. So what happens to the copper and associated minerals when basalt is altered from magnetite to hematite by extremely hot, mineral-rich waters?

Haynes says other geologists had previously noted an association between altered basalts and copper occurrences, but there was no proof that altered basalts produced copper. The PhD study sought to prove the connection. If Haynes succeeded, he would establish for Western Mining a radically different model for finding copper deposits. It was exactly the kind of new science on which the company had built its reputation and its rising wealth.

Woodall was excited by the idea and encouraged the choice of topic. Haynes says Woodall also made a critically important contribution by insisting the PhD was undertaken at a university of international standing. "It was through Roy that I went to The Australian National University in Canberra. ANU was the premier earth science department in Australia, and the most international in flavour. It was an incredibly good decision to go there," Haynes says.

He took study leave from the company and began his PhD in May 1969, under the guidance of Professor Allan White and others in the ANU's Department of Geology. Haynes wrote in 2006 that White "must surely be one of the more inspirational mentors of undergraduates and graduate students in the Australian university research and teaching scene in the 1970s and 1980s.

"His continued emphasis on the need to ask thoughtful, but not necessarily 'correct' scientific questions, to think laterally, to not accept the conventional, and especially, his encouragement of new ideas provided outstanding guidance to a young PhD student. Allan's enthusiasm made the science more enjoyable, and his mentoring dramatically improved the quality of the research.[8]"

Studying at ANU also brought Haynes into close contact with Bruce Chappell, who was among a very select group of geochemists invited by NASA to study basalts recently brought back from the Moon. Chappell's expertise was world-class, as was his ANU laboratory. Both would be available for the next three years to help the young Haynes tackle the more earthly mysteries of the Warburton basalts.

Haynes' expenses during his PhD study were covered by a Government-funded post-graduate scholarship. Western Mining also contributed by meeting all the costs of fieldwork during the PhD, including flights to and from the Warburton Ranges.

"Western Mining's study leave was a marvellous scheme. They would keep your superannuation intact, keep your staff position, pay your airfares and initial costs of getting into accommodation. That was a terrific scheme, way ahead of its time," Haynes says.

---

[8] Mentoring and the Olympic Dam Ore Deposit Discovery - A Personal View, by Douglas Haynes, 2006, presented at Australian Society of Exploration Geophysicists-Geoscience Australia Earth conference

Half-way through the PhD program, Haynes became the father of twins. "It was pretty hard to do the research, so I took a couple of months off just trying to help my wife, Estelle, manage. But the Commonwealth scholarship had pretty rigid conditions, they wouldn't extend the scholarship. So Roy Woodall stepped in and paid half my salary for the additional three months I took to finish it."

He says support like that was typical of the company in those days, and reflected a very strong sense of camaraderie in the Exploration Division. "It really did feel like a family. We all had a great feeling of enthusiasm for the company and were totally in this spirit of wanting our company to be successful."

Haynes says that by early 1972 — almost three years into the PhD study — the research showed convincingly that common types of continental basalts became "potent sources of copper" when altered by heated water in the Earth's crust. Furthermore, certain chemical markers (such as the ratio of reduced and oxidised iron in the rock and the rock sodium content) clearly identified some basalts as better sources of copper than others.

Haynes says that, as these findings became clear, he had many interesting discussions with Roy Woodall and Jim Lalor, the future leader of the search for Olympic Dam, about what they might mean for copper exploration in Australia.

Haynes' PhD received a commendation from the ANU review committee, falling short of the top level of merit. Perhaps the ANU did not appreciate the full significance of the research, although it would rectify this in 1985 by awarding Haynes the David Brown Medal for the best application of a PhD to industry.

Western Mining certainly had no doubts about the value of Haynes' findings. Haynes says everything possible was done to keep his research from the prying eyes of competitors. While there were limits to the restrictions that could be placed on access to a taxpayer-funded PhD, Western Mining was able to keep Haynes' research findings under wraps until 1974. It effectively managed a two-year head-start on any competitors that might want to throw their own exploration funds behind the breakthrough science that Haynes had developed at ANU.

In August 1972, Haynes and his newly enlarged family up-rooted from Canberra and travelled back across the country to resume their life as part of Western Mining's Exploration Division in Kalgoorlie. Woodall appointed Haynes, still only 28 years of age, to a new role of "Copper Consultant" and gave him the autonomy to operate however he saw fit. Haynes says he believes this freedom was crucial in finding a pathway that would lead to the discovery of Olympic Dam.

The search begins

Haynes' discovery revolutionised Western Mining's search for copper because the nature of the target was now dramatically different. The company was no longer searching for narrow and easily-missed veins of copper, as it had done at Warburton and Moonta. Nor was it searching for sedimentary rocks in which copper had been deposited, a concept already pursued in the Hamersley Ranges without success for five years.

Now it wanted altered basalt — a reddish, heavy and fine-grained volcanic rock. The company's geological leaders knew that if they could find altered basalt — the potential source of the copper — the prize of an ore body should not be far away.

Altered basalt as an exploration target has a number of advantages. It has distinctive physical properties, which would later become crucial in the search. But more importantly in the initial exploration stage, altered basalt is easy to recognise, even by untrained eyes. Young basalt from lava floodplains in Victoria was heavily quarried and is the bluestone that practically defines the cityscape of Melbourne. Basalt is well recorded on maps just about anywhere it is found.

By contrast, the types of rocks to host sedimentary copper deposits are hard to distinguish from many other sedimentary rock formations. Identification is made even more difficult by their softness, old age and heavily weathered appearance.

Haynes began the search for new altered basalt targets with a national perspective; this was an entirely new exploration model and the whole country was effectively a blank sheet of paper.

From his office in Kalgoorlie, he acquired every map and geological record he could find with occurrences of altered basalt. Much of this data came from the Bureau of Mineral Resources, established by the Federal Government in 1946 to systematically map the geology and geophysics of Australia. The BMR is a forerunner of Geoscience Australia. Equivalent state government organisations, known as State Geological Surveys, also provided material for the new search. Some of these had compiled for more than a century large amounts of information that would be useful to Western Mining's innovative search for copper.

According to the new exploration model, sedimentary rocks would contain copper that had leached from an underlying, altered basalt. If the sedimentary rocks outcropped, either prospectors had already found the copper or it was not there in the first place. In either case, Haynes could rule out the location. But if overlying rocks were concealed, there was a strong possibility they contained a copper deposit. The only way to answer the question was to drill an exploration hole.

The absence of copper in sedimentary rocks above an altered basalt does not disprove Haynes' PhD or invalidate the new copper exploration model. The sedimentary rocks may not have the necessary faults or cracks to provide a pathway for the copper. Or perhaps the copper made the journey, but the chemical conditions in the sedimentary rocks were not right to bring it out of its dissolved state and deposit the copper minerals. There are many reasons a predicted copper deposit could be missing, which is part of the uncertainty and the thrill of mineral exploration.

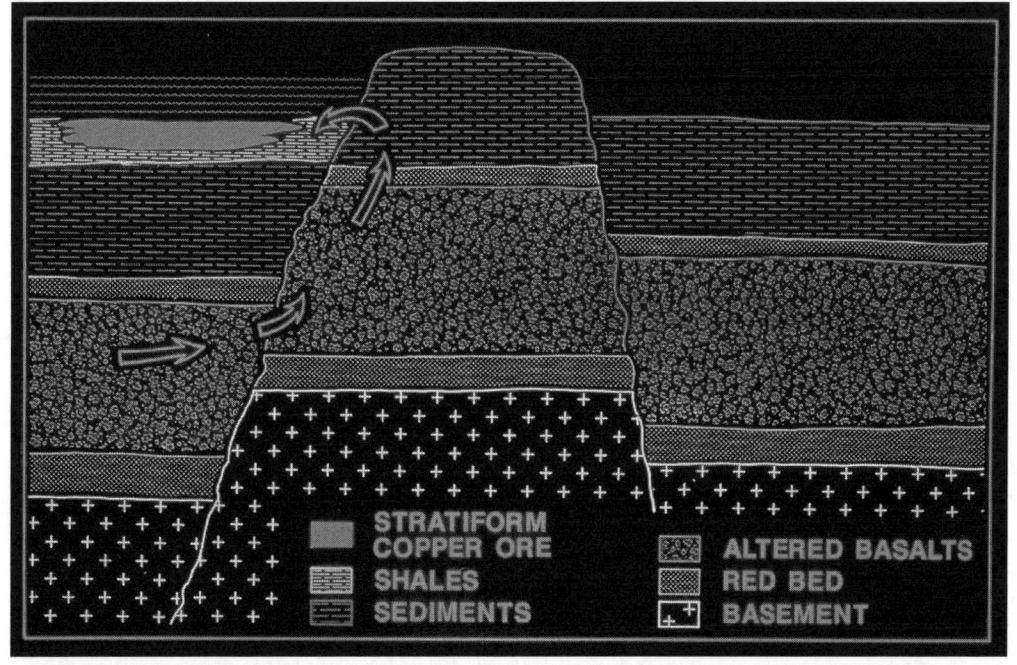

A graphic of Western Mining's altered basalt model for stratiform copper deposits, from a company presentation several years after the discovery.

Haynes' preferred altered basalts were old rocks of Proterozoic age, spanning a period between 2,000 million and 545 million years ago. This era was chosen because Proterozoic-age altered basalts are associated with many of the world's great copper deposits. The Warburton Range basalts Haynes studied were in the middle of this age range at about 1,100 million years old, and the copper deposits of Mt Isa, which were then the largest in Australia, were of a similar age.

If Haynes was really lucky, government geologists might also be able to tell him whether the Proterozoic-age basalts on their maps were altered by hydrothermal fluids, which was necessary for them to become a source of copper. This kind of detailed chemical analysis was unusual at the time, but Haynes had some good fortune. The BMR had published an extensive compilation of the chemical composition of basalts and other igneous rocks around Australia by Germaine Joplin, one of the female pioneers of geology in Australia in the 1920s and 30s.

Haynes says this information was a good source and helped to narrow the search, although it was still an arduous task over several months. It was not until late 1972 or early 1973, that Haynes had compiled a national map of prospective locations for copper exploration.

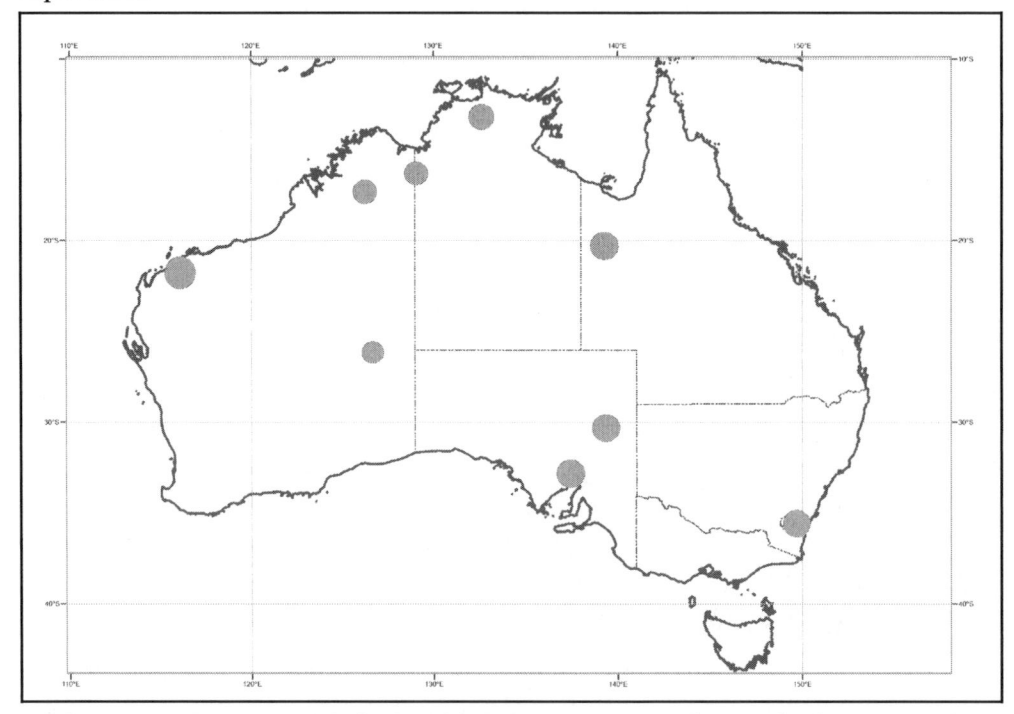

The pain-staking research had yielded five targets — the Kimberleys in northern Western Australia, the northwest Northern Territory, northwest Queensland, southern New South Wales and several areas in South Australia. All of these regions had altered basalt of Proterozoic age, but there was one more important criteria in their selection; the basalts in all these areas were covered by sedimentary rocks.

Haynes says the search produced so many possible locations around Australia that a decision had to made about which area to tackle first. "So, I asked Roy. He said: 'Start in South Australia. They have a progressive Department of Mines and Western Mining has a good standing with them because of our exploration activity at Moonta over many years.'

"I don't recall him talking about selecting South Australia because of its great history of copper occurrences, but he probably did. Roy always emphasised the importance of following up on the presence of small indicator mineral occurences in successful mineral exploration. That was the advice from Roy Woodall, which was incredibly prescient when you think about it. If we had started in northwest Queensland, which was number two on our list, maybe we would have found copper deposits up there, but certainly Olympic Dam would not have been found at the time it was."

And so Western Mining began closing in on Olympic Dam.

Opposite page: Early target areas for altered basalts, identified by Haynes in 1972. The map shows five broad target areas plus Warburton near the junction of Western Australia, South Australia and the Northern Territory. Western Mining's limited resources meant the focus had to be narrowed to one state. South Australia was chosen first. Map source: Source: Douglas Haynes' presentation to the 2006 Earth conference organised by the Australian Society of Exploration Geophysicists and Geoscience Australia.

7

HOMING IN ON ANDAMOOKA

Douglas Haynes' next field trip would be two months in the Travelodge on Adelaide's South Terrace. The new destination hardly compared with the isolated and ruggedly beautiful places he studied in the Warburton Ranges and the Hamersleys, but that's where the search for copper led in early 1973.

Now that South Australia was the focus of the search, it was time to make an exhaustive study of the state's Proterozoic-age altered basalts. South Australia's Department of Mines — just across Adelaide's parklands from the Travelodge — held many reports on exploration drill cores and outcrops containing basalt.

Government geologists and private explorers had compiled these records over nearly a century of exploration of the state's mineral wealth. Private explorers were encouraged by the department to place on public or 'open' file the results of unsuccessful exploration. Given that around 95% of all exploration is unsuccessful, a huge volume of data was now available.

Of course, previous explorers did not understand the importance of altered basalts in the formation of copper deposits, or how they might lead an explorer to an orebody. Western Mining's new knowledge meant the public files — rich with information on altered basalt from exploration with other goals in mind — might now have immense value. Unfortunately for Haynes, there are hundreds of occurrences of basalt in South Australia and the public records seemed to contain information on most of them. A painstaking search of the files was the only way to find the valuable clues.

Haynes says he spent a total of about two months based at the Travelodge, breaking the stretch with a number of trips back home to Kalgoorlie. But many consecutive days began with a walk across the parklands, followed by long hours immersed in piles of records, and the walk back to his hotel at the end of the day.

"It would have been a couple of months, working through all the old open-file exploration reports. And bear in mind that in those times we didn't have microfiche, just horrible paper copies. You had sit down in this dusty room and wade through all this bumpf, trying to distill the essence. It was pretty tough work, pretty boring."

The hard slog was paying off. Haynes' had found more evidence of the right types of altered basalts, and the broader geological setting was also encouraging. Under its new model, Western Mining wanted deep basalts that were depleted in copper, buried under sediments that were in turn hidden from prospectors by younger sediments. The final ingredient in the model — faults or cracks to provide a pathway for the copper into the overlying sediments — also seemed to be present in a number of locations.

Roy Woodall was ready to take the search up a gear. The most pressing need was for an exploration base in Adelaide, which Western Mining did not have despite many years of exploration in the Moonta-Wallaroo region, about 150 kilometres to the north.

Woodall asked Dan Evans, a young American geologist who joined Western Mining in 1971, if he was interested in setting up an exploration office in Adelaide. It would be an office of only one professional and one part-time assistant, but 100% of Evans' time would be committed to the search for copper under Haynes' new model.

Evans was recruited by Jim Lalor, who recently returned from study leave at University of Manitoba, Winnipeg. Lalor

Dan Evans near Kalgoorlie in 1972

was interested in exploring for copper-zinc massive sulphide deposits in Western Australia. Evans was completing a Masters Degree at McGill University, Montreal and had a deep understanding of the copper-zinc massive sulphide deposits in Canada's Archean-aged rocks. Lalor believed these skills might unearth similar deposits in the Archean rocks of the Yilgarn Craton in Western Australia. Two years of exploration, first under Lalor and then Woodall, produced very limited success and showed that the analogy between Australian and Canada did not easily apply. With his assignment complete, Evans and his Canadian wife, Linda, moved to Kalgoorlie, 700 kilometres south of their home for the past two years in the hot and isolated town of Meekatharra.

The young couple had passed through Adelaide on their arrival in Australia. It struck them as a great place to live that could offer a complementary experience to life in outback. Their original plans to return to Canada after a couple of years were put on hold for the opportunity to take up a more senior role in an attractive city.

Before moving to Adelaide, Evans would have a few months in the Kalgoorlie office, where he got to know Haynes and his new model for finding copper. Haynes says Evans quickly became an enthusiastic supporter of the search and its aims.

In February 1974, the Evans family relocated to a rented home in Flagstaff Hill, a new southern suburb less than 30 minutes from the city centre. It was a comfortable home with every amenity of the day, including a two-car garage with a separate entrance and even its own toilet. But the Evans family would be parking in the driveway — the garage was reserved as the first Adelaide office of Western Mining.

The Daveys Road home is still there and arguably deserves more recognition for its role in one of the world's greatest mineral discoveries. From this garage, the search for copper was narrowed from the whole of South Australia to Olympic Dam. Evans, Haynes and a growing team of Western Mining's best geoscientists met in the garage — sometimes in groups of up to seven people at a time. From here they hatched much of their exploration plans and conducted dozens of field trips. Drill core from the first two exploration holes was even stored in the backyard, which became the first coreyard in which Olympic Dam drillholes were logged. This would later cause a brief flurry of public controversy because of concerns about the safety of the uranium minerals.

The company's low-budget approach to setting up an exploration base seems at odds with modern-day perceptions of Western Mining as a multi-billion dollar enterprise, but it reflected the tough times of the early 1970s. The world had been dragged into economic recession by the oil price shock, the mounting costs of the Vietnam War and other economic factors. The price of all metals fell sharply, including nickel, Western Mining's major profit earner since 1968. The severity of the downturn is highlighted by collapse in Western Mining's market capitalisation from more than $4 billion in current dollar terms ($484 million in 1972 dollars) at the height of the nickel boom in 1972 to only $800 million ($140 million) in 1975. (Market capitalisation is a measure of the value of a company listed on the stock exchange. It is calculated by multiplying a company's share price by the total number of shares held by investors).

In these times, companies of all sizes were firing staff, cutting budgets and shelving growth plans. It was against this background that Western Mining stepped up its copper exploration program and established an office in Adelaide. The fact that it found a low-cost way to do this says much about its commitment to exploration as well as its ingenuity. Decades of lean times had taught it the art of mineral discovery on impossibly small budgets.

By the time Evans arrived in Adelaide in early 1974, Haynes had been through all the publicly available exploration data. With Evans now on the ground, the company had its own geologist to gather new information from first-hand observation in the field. Altered basalt occurrences identified from Haynes' work were a major focus, but Evans covered much more ground. His most fundamental objective was to provide Western Mining with a first-hand view of South Australian geology. As a rule, the company did not put its faith in the geological observations and mapping of others; Evans would build a base of knowledge the company could trust by preparing a series of geological orientation reports on the state. This knowledge and Evans' experience of the state's geology would become crucial in narrowing the range of possible exploration targets.

Evans and Haynes also spent time in the early months of 1974 in the core library of the Department of Mines, in the inner-eastern suburb of Glenside. The Department had recently established a central archive to store the physical records of government-fund core drilling. The archive also included large amounts of exploration core from the private sector, which had jumped at the government's invitation to provide a central storage facility of core from exploration activity that was now on the public files.

Haynes and Evans were given permission to take samples from cores in the public archive. They were particularly interested in those that showed alteration from magnetite to hematite, which was a primary indicator of copper depletion.

Geophysics enters the picture

The new search for copper in South Australia was being escalated on a number of fronts. The decision in late 1973 to establish an exploration base in Adelaide was a big practical step and brought Dan Evans into the team. At the same time, Woodall and Haynes started to reach out to experts in other geoscience disciplines in Western Mining. The most important of these was geophysics. In a conventional copper exploration program, geologists carry most of the workload. They can be out in the field for years, as they were in the Hamersleys and at Tarraji River, mapping rock formations and geological structures, and taking thousands of soil and rock samples to test for copper. But under Haynes' new copper exploration model, the most important rocks of interest were buried. This problem was always in Haynes' mind when he developed the model, but the distinctive physical properties of basalt would once again provide a solution.

Basalt not only looks different to most rocks, it has physical properties that can indicate its location deep below the surface without any exploration drilling. Basalt is dense, weighing between 2.7 and 3.3 grams per cubic centimetre. It can be up to 30% heavier than granite, which in turn is heavier than many sedimentary rocks. This difference is large enough to increase the force of gravity measured at the surface above a body of deeply buried basalt. Basalts are also magnetic because they typically contain a significant percentage of magnetite, an iron oxide and the most magnetic of all naturally occurring minerals. The magnetic field generated by basalt can be so strong that an ordinary mariner's compass was used as early as 1640 to find magnetite ore in Sweden[9].

By systematically measuring gravity and magnetic forces across a wide area, geophysicists can identify locations with unusual or anomalously high readings. These could suggest where a body of basalt might lie at depths of hundreds and even thousands of metres below the surface. Geophysical surveys cannot prove a basalt occurrence, but if a gravity "high" and a magnetic "high" are found in the same location, there is certainly a large amount of dense, magnetic rock below. If the regional geological setting is right, the odds of these deep, dense rocks being basalt are further improved.

In the new search for copper, geophysics would be vital. Haynes' geology had created the opportunity, but geophysics would be heavily relied upon to pinpoint the target for exploration drilling.

In late 1973, Haynes explained his new copper exploration model to Hugh Rutter, a 32-year-old Englishman in charge of all of the company's geophysical exploration activity outside Western Australia. Rutter joined Western Mining in 1969, six years after graduating with a geology degree from Durham University in the north of England. The career prospects for geologists in the United Kingdom were unexciting, so Rutter and his wife, Anthea, moved in 1967 to Perth, the centre of a nickel boom that was making headlines around the world. The plan was to make some good money and return to the UK after a couple of years, but this was quickly overtaken by Rutter's success in the wide, open spaces of Australia.

---

[9] Carlsborg H., 1963. Om gruvkompaser, malmletning och kompassgangare. *Med Hammare och Fackla* (Stockholm), 1963(23):9-10

After 18 months working for the Geological Survey of Western Australia, Rutter was recruited by Western Mining's chief geophysicist, Anton Triglavcanin. His new role in the Kalgoorlie office was to apply geophysics in the search for more of the nickel sulphide deposits that had recently transformed Western Mining. He was spectacularly successful, thanks to his expertise in some quirky Russian technology.

"Western Mining had recently brought in a technology new to Australia, Time Domain Electromagnetics. We were using a Russian instrument known as an MPP01. We did this little experiment with it at a small nickel ore body at Kambalda, called the Edwin Shoot. Doing a test line, I found that it showed a response over a known nickel ore shoot. I extended the test line a bit further to get some background readings, and unexpectedly found another response. I pointed it out to Roy Woodall, who said: 'What's that?' I said it might be another nickel shoot. 'Well, we'll get a drill down there. There's one spare tomorrow.' Of course, my heart sank. I was being put to the test! But we drilled on this anomaly and it was another nickel ore shoot, which became Edwin East. After that, I was in the good books."

Rutter was selected for Western Mining's study leave program, which meant the opportunity to do post-graduate study while on half-pay. He seized the chance to go back to the UK to study for a Masters Degree in geophysics at the Royal School of Mines, part of the University of London. In typical Western Mining style, he was being sent to an institution where he would be exposed to the very best thinking in the world.

Soon after returning to Australia, Rutter was promoted to take charge of geophysics in the eastern states. It was a big promotion, but Woodall as always was eager to advance people who had used the best science to make discoveries. Rutter was now part of the Eastern Australia Exploration Division, which ran all exploration outside Western Australia from a former single-fronted shop in the Melbourne suburb of Preston.

Hugh Rutter in 1978

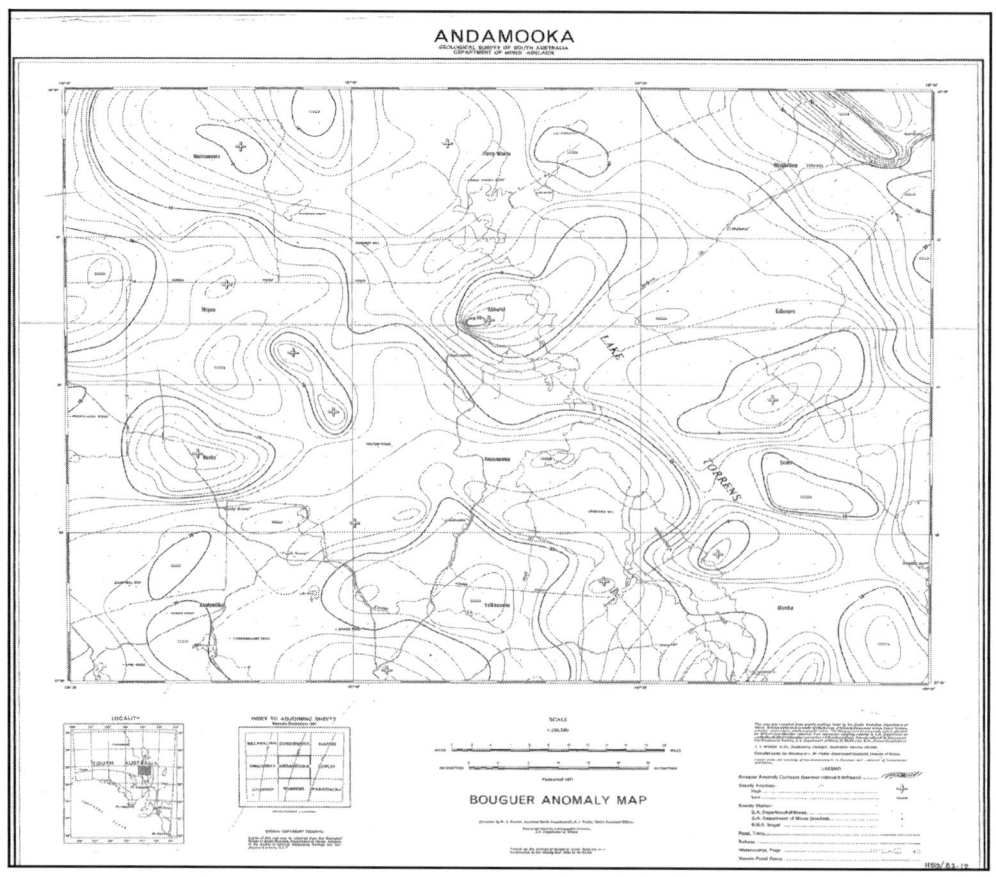

Geophysical maps show the distribution of natural gravitational and magnetic forces at the Earth's surface. The swirling lines connect points of equal force, in the same way contours on a topographical map show points of equal elevation. The maps reveal areas of high and low gravity or magnetics, known as anomalies, which point the way to deeply buried rocks of possible interest to a mineral explorer. In the 1970s, maps such as this gravity map of the Andamooka region from the archive of Geoscience Australia were contoured and drawn by hand, and required almost as much art as science on the part of the geophysicist. To a layperson, the anomalies look like random patterns, but a geophysicist sees potential meaning in every curve and gradient. Before computers, interpretation of maps required weeks of manual calculations to try to fit the anomalies to possible configurations of subsurface rock units of various shape, orientation and density. Many permutations can be made to fit any given anomaly, which is why some regarded geophysical interpretation as something of a black art. Today, computer modelling has removed the mystique.

Tectonic lineament studies

Western Mining had one other geoscience to help in the search for "blind" copper deposits — a controversial branch of structural geology known as tectonic lineaments. The company's tectonic lineament studies were led by Tim O'Driscoll, who originally joined Western Mining in 1939 just six years after W.S. Robinson founded the company. In 1942, O'Driscoll left to serve in the RAAF. He rejoined the company in 1970 after working in academic posts, the private sector and government, including a number of years as chief geologist in the South Australian Geological Survey.

O'Driscoll, who died in 2004, was fascinated by big structures in the Earth's crust and how they were expressed in the landscape. In particular, he wanted to understand how landscape features might point the way to ore deposits. Woodall is also a strong believer in the importance of deep Earth fractures, expressed as linear discontinuities in geological and geophysical data, known as lineaments. He recruited O'Driscoll and a team of structural geologists under him to map tectonic lineaments across Australia. An office was established for this group in a sub-let area of the Australian Mineral Development Laboratories in Glenside, Adelaide, shortly after Evans had established the exploration base at Flagstaff Hill.

Lineaments are best explained in *Crustal Structures and Mineral Deposits*[10], published in 2007 as a tribute to O'Driscoll's life and work. "Often, when gazing at a map, but thinking about something else or nothing, lines magically appear in the mind's eye. For instance, sitting in a lecture theatre waiting for proceedings to begin, a wall map of the world or of a continent or region suddenly comes alive with straight coastal sectors lining up in parallel with each other or with rivers, or mountain fronts continuous with the margins of lakes and so on. Such involuntary imposition of lines on maps is seductive and possibly meaningless, just as one occasionally and ephemerally discerns faces and other

Tim O'Driscoll

[10] *Crustal Structures and Mineral Deposits, E.S.T. O'Driscoll's Contribution to Mineral Exploration,* 2007, Rosenberg Publishing, copyright the O'Driscoll Volume Committee

objects in patterns of wallpaper and so on; but the cartographic alignments referred to are real, for the continents are indeed criss-crossed by linear features which are the surface expression of crustal structures."

O'Driscoll's work at Kalgoorlie and Broken Hill led him to advance the idea that orebodies were often located at the intersection of lineaments and the deep crustal structures they represented. He theorised that an intersection of lineaments provided a pathway for the migration of hydrothermal fluids from deep in the crust to shallower zones, where a change in pressure and temperature could lead to the formation of an ore deposit. This type of thinking could be very important in the new model for copper

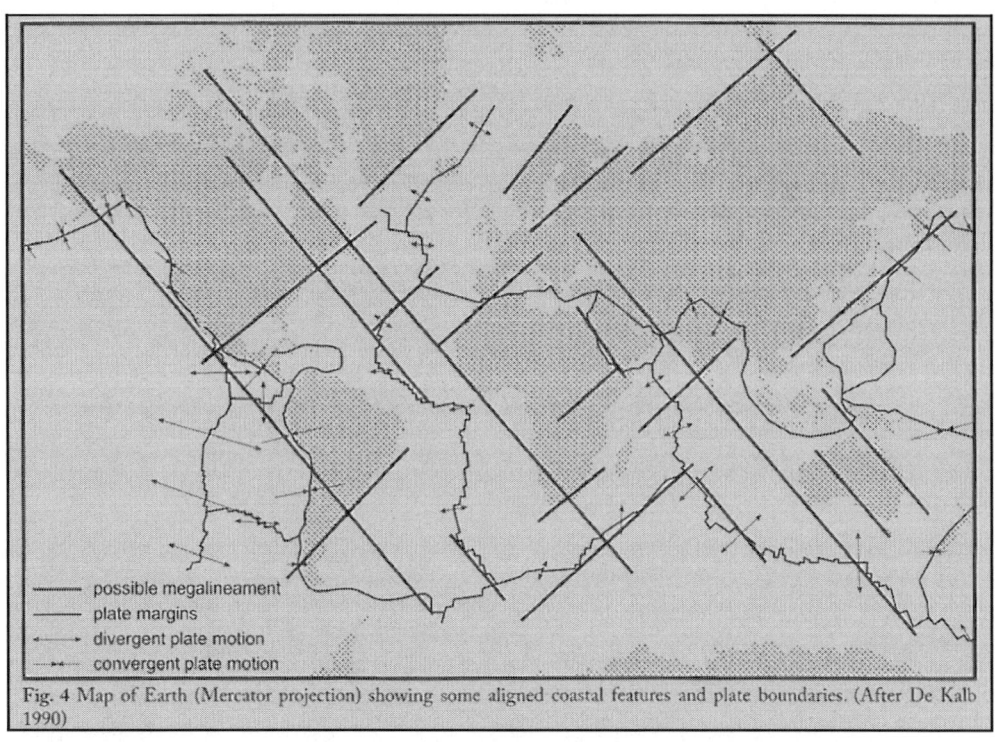

Fig. 4 Map of Earth (Mercator projection) showing some aligned coastal features and plate boundaries. (After De Kalb 1990)

Above: A map from Crustal Structures and Mineral Deposits. Opposite page: O'Driscoll spent weeks studying aerial photos of the Andamooka mapsheet to produce this tectonic lineament study, which is from the BHP Billiton Archive. The thousands of short lines were traces of vegetation patterns. These were used to interpret the position and orientation of the sets of large lines. Some of the intersection points were circled to indicate tectonic targets.

exploration because some kind of structural pathway was needed for the copper to travel from altered basalts to overlying sediments.

The study of tectonic lineaments remains controversial and is still not accepted by many geologists. Those who believe in the study say linear features in the landscape reflect underlying geological structures. Critics of lineaments argue that some features are merely random artifacts of nature and it is not possible to distinguish these from lineaments that reflect important structures in the Earth's crust. Woodall says the geoscientific community is slow to accept major new ideas. He states the study of lineaments and their value in understanding crustal structures will one day rank alongside plate tectonics as one the most important theories in geoscience.

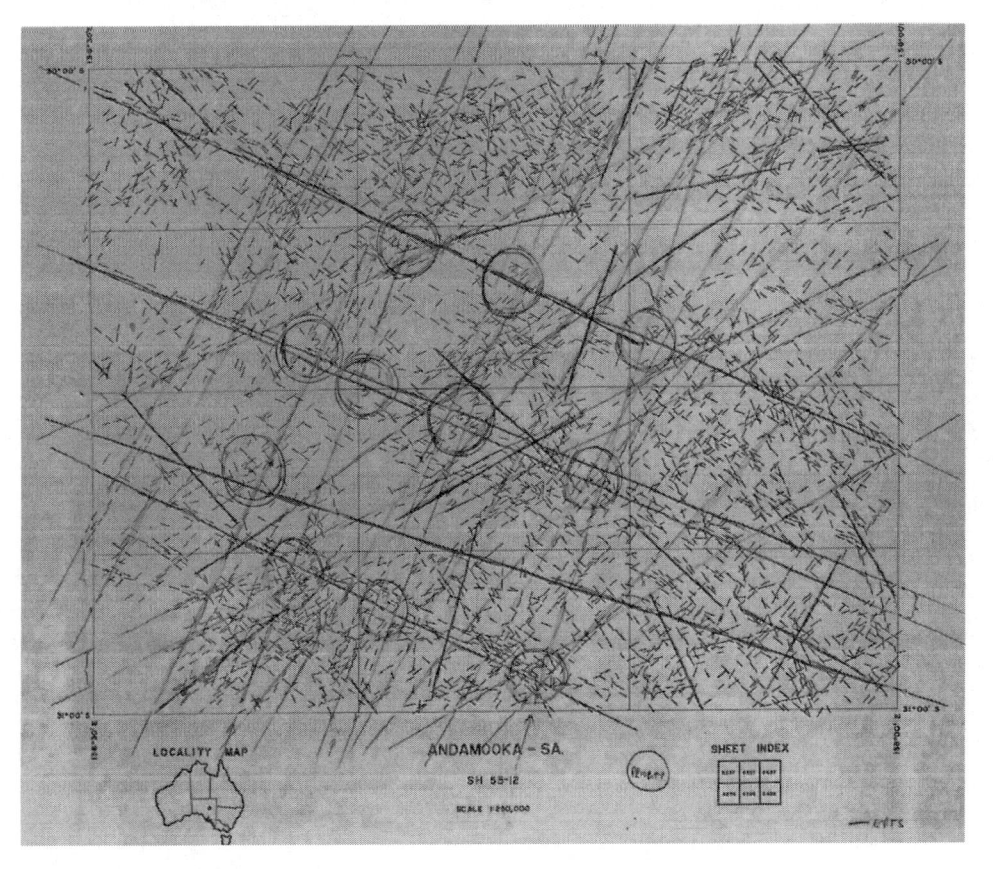

The cause of lineament studies is not helped by its followers' claims they can identify linear features in the most unlikely places, even in the monotonous terrain of sand dunes and clay pans around Olympic Dam. How can deep structures in the Earth's crust be identified beneath a cover of shifting sand and mud? O'Driscoll believed the distribution of vegetation could reveal lineaments. He argued that cracks in rock beds below the sand dunes provided plants with the best access to groundwater. The distribution of vegetation therefore reflected rock fracture patterns, which were created by constant movements above the location of large-scale lineaments.

Finding the vegetation and rock fracture patterns involves weeks of painstaking study of aerial photos. The map on the previous page shows thousands of fractures that O'Driscoll and his team interpreted from studying vegetation in aerial photos of the Olympic Dam area. These hand-drawn lineament studies were often referred to as "chicken track" drawings. For some it was an affectionate reference to the intricate patterns of lines, but for others it seemed an appropriate label for work they considered to be mystifying.

The new team leader

New people with new skills were now being applied to the search. As with any multi-disciplinary team, the potential for conflict and chaos was high unless a strong leader emerged to take control. The new Adelaide exploration office was part of the Eastern Australia Exploration Division, so divisional head, Jim Lalor, assumed responsibility for the new and highly experimental search for copper in South Australia. Lalor was still relatively young at 37 years of age, but by all accounts a great and instinctive leader, a trait he may have inherited from Peter Lalor, who famously led the protests of gold miners at Ballarat in 1854. (Jim Lalor's family has reason to believe Peter was a relative, although documentary proof has always been elusive).

Lalor joined Western Mining subsidiary Gold Mines of Kalgoorlie as soon as he finished an Honours year studying Proterozoic sediments at The University of Western Australia. He had planned to continue in academia to undertake a PhD and eventually join the resources industry as a

Jim Lalor in the late 1980s

petroleum geologist. But at the end of his Honours year in 1959 an offer from Western Mining to become an underground geologist in Kalgoorlie was too good to refuse. He did not take long to make an impact. In 1961, Lalor and David Barr discovered the Koolanooka Hills iron ore deposit, just inland from Geraldton, Western Australia. Lalor had seen iron-rich sediments working in the region during his Honours year. Koolanooka Hills was a small mine by today's standards, but it was the first in Australia to export iron ore following the lifting of a decades-old ban. The Australian Government had prohibited iron ore exports because it was believed Australia's limited resources should be reserved for the country's own steel mills.

In 2006, Haynes wrote[11] about the importance of Lalor and Evans in the newly formed team. "Dan's input at this stage of the project was morale boosting and very important, with great support given through his sheer enthusiasm coupled with unstinting support in use of the exploration base facilities during the various field reconnaissance spells, and through numerous discussions presenting alternative points of view, which improved the operational procedures utilised in the project. Even at this relatively early stage of the program, Jim's and Dan's support was especially important in dispelling some of my strong uncertainty and doubts attending application of an untried copper deposit targeting model, especially because emphasis was to be on exploration for blind or concealed copper mineralisation."

Homing in on Andamooka

The search rapidly gained momentum. In the first year after Haynes returned from his PhD study in August 1972, South Australia was identified as the first priority in Western Mining's new quest for copper, while Evans knew he would be off to Adelaide to establish an exploration base in the New Year and the first steps were being made to assemble a multi-disciplinary team. Over the next year, the search would narrow to a region around the opal mining town of Andamooka in the far north of the state.

The following chronology of events over this period has been compiled from Western Mining's K/2792 report by Lalor in 1986, supplemented by various articles and presentations by team members, and by interviews conducted for this book. The K/2792 report, which extends to more than 160 pages, was prepared a decade after the discovery with the aim of creating a definitive record. It compiled all relevant written records within Western Mining from 1972 to 1976, although some team members point out it is still incomplete because the records did not always keep up with the fast pace of events.

---

[11] Mentoring and the Olympic Dam Ore Deposit discovery - a personal view, 2006 presentation to ASEG-GSA Earth conference

The report highlights better than any other document the breadth of the search. Many accounts of the discovery create the impression that Western Mining's geoscientists were drawn easily to Olympic Dam. The team was first attracted in February 1974 to drilling geophysical targets around Andamooka, but it did not settle on this idea for another six to eight months. In the meantime, there was discussion of drilling near Port Pirie and other targets just north of Port Augusta. In May 1974, the team was still considering nine very large areas, scattered across almost 300,000 square kilometres or one third of the state. Narrowing the search from this vast area to Olympic Dam target required thousands of hours of effort.

Lalor's record also highlights the plague of uncertainty. Like all other mineral explorers, the team members knew they had to work hard to overcome the high odds of failure. They certainly did not know they would succeed. Optimism and faith in the science behind their endeavours were far more important than we can readily appreciate today with the knowledge that a great discovery awaited the team. The report also dispels a perception the team set out to be unconventional almost for its own sake. It's true they wanted an exploration model that could take them to places not already covered by prospectors and competitors, but science was always the ultimate guide of their search. Many of the areas they were taken to by the altered basalt model had already been well explored for copper, some were even under active exploration by competitors, although not yet covered by an exploration licence. This created some interesting challenges, as revealed by Haynes in a memo to the rest of the team in June 1974 about exploration plans for Wooltana, about 300 kilometres east of Olympic Dam on the other side of the Flinders Ranges.

Wooltana was a highly favoured target in early 1974 after a two-metre thick exposure of rocks stained with malachite (copper carbonate) was found where the team had predicted in a rough gorge. Selection Trust had explored around Wooltana several years earlier and Anglo American was now taking a close look, along with Western Mining. Haynes wrote: "Because Anglo American are interested in this area, we should lie doggo — no soil sampling or rock chip sampling should be done at this stage. However, B. Severne (a specialist retained by Western Mining in geobotany or the science of identifying minerals from the type of local vegetation) disguised in a butterfly net could systematically traverse horizons of interest looking for indicator plants."

The memo reveals Haynes' humour, but his message was serious and says much about the hunt for Olympic Dam in early 1970s. Haynes and others in the team were fiercely competitive and determined to employ any tool that might improve the odds of success. Haynes talks today about a culture within Exploration Division of wanting to succeed for Western Mining, and this is evident in many of the written records of the search.

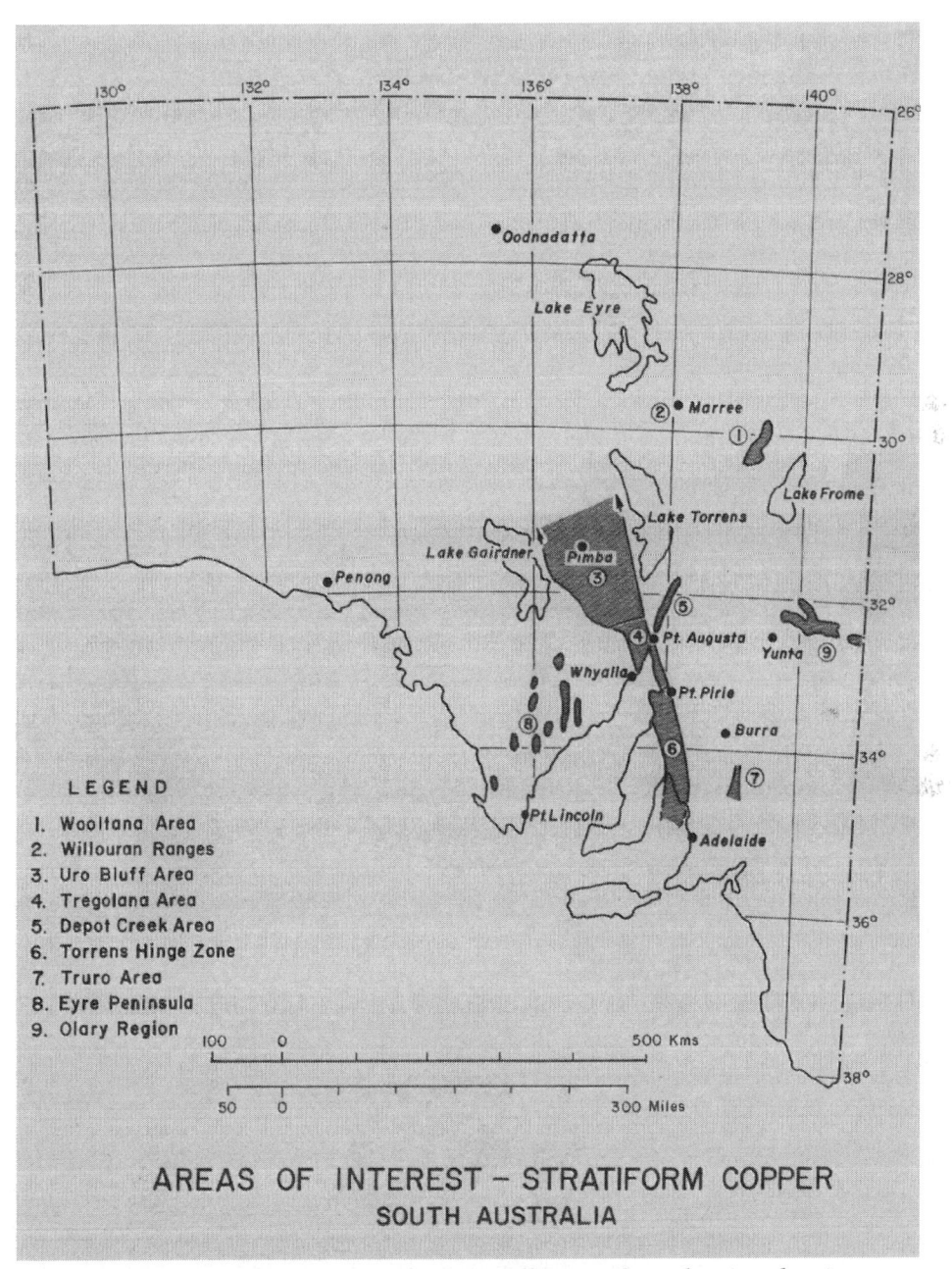

A map from the K/2792 report in the BHP Billiton archive showing the nine target areas in early 1974. Andamooka was not a target at this stage, but the team was closing in. Tregolana (area 4) included the Roopena basalts.

The team relied on three or four critical pieces of information to narrow the target to Olympic Dam. The first of these was the discovery at Roopena Station, immediately northwest of Whyalla, of precisely the kind of altered basalts targeted by the national search that began back in late 1972. These rocks had the correct chemical composition and had lost up to 60% of their original copper content. Roopena is a long way from Olympic Dam, almost 350 kilometres south of where they would eventually make their discovery. Furthermore, a huge obstacle known as the Stuart Shelf lay between these two locations of interest.

The Stuart Shelf is a regional or large-scale geological feature that extends from north of Whyalla almost as far as Lake Eyre. As the name suggests, the Stuart Shelf is a shelf or platform of sedimentary rocks, known collectively as the Adelaidean group, that overlie the older Gawler Craton. Sediments poured onto the Gawler Craton while also filling a deep trough extending from south of Adelaide to the northern tip of the Flinders Ranges. The trough-bound sediments were uplifted to form the Mt Lofty and Flinders Ranges, but the sediments that spilled over the Gawler Craton have sat flat and largely undisturbed for up to 1,000 million years. The Roopena basalts pre-date the oldest of these sediments, which means they would be hidden under Stuart Shelf, if indeed they did extend north. There were no outcrops of deeper rocks for hundreds of kilometres and the team did not even have drilling results from other explorers to provide clues. There had been no drilling below 100 metres anywhere on the Stuart Shelf except for three scattered water bores. Despite all these obstacles, the team uncovered pieces of evidence to suggest altered, copper-depleted basalts extended from Whyalla, where the Gawler Craton was exposed, all the way to Andamooka and beyond.

The first clue came from an old bore at Beda Hill, about 70 kilometres north of Port Augusta and 140 kilometres north of Roopena. South Australia's Public Works Department sunk the bore between 1888 and 1890, looking for water along a stock route from Beda to Andamooka, according to South Australian geologist, David Tonkin[12]. The bore was dry, but drill cuttings were sampled and stored for almost a century in an Adelaide basement belonging to the Engineering and Water Supply. When the E&WS moved in 1974, the samples were rediscovered and for a while in danger of being thrown out. Thankfully, the Department of Mines was given the opportunity to take custody of the samples in its drill core library, where Haynes and Evans found them as part of their exhaustive search of the state's records of altered basalts. Chemical analysis of these rocks at Western Mining's laboratory in Ballarat

---

[12] *Seeking the Beda Bore*, MESA Journal, December 2009, by David Tonkin

found they echoed the characteristics of the rocks at Roopena. (They would later be identified as a distinct unit known as the Beda Volcanics).

The great thickness of basalt intersected by the Beda bore also added to the team's confidence in the idea of extensive sheets or piles of copper-depleted basalt under the sediments of the Stuart Shelf. The Beda bore intersected basalt at 140 metres and was still in basalt when the bore was abandoned at a depth of 335 metres after two years of drilling. The persistence of the Public Works Department at Beda was remarkable and fortunate. Tonkin writes that Port Augusta residents held public protests in 1890 in a bid to continue drilling at the bore. Whatever their motivation, the protestors almost certainly kept the bore going deeper than planned and helped Western Mining's geoscientists track down Olympic Dam almost 90 years later.

Another critical clue was the Mt Gunson copper mine, another 60 kilometres north of Beda Hill. Copper ore had been mined by hand at Mt Gunson as early as 1898. Open-cut mining had resumed in the early 1970s after the discovery of the Cattle Grid deposit, a small, flat-lying orebody at a depth of only 30 metres. Cattle Grid was a stratiform copper deposit, similar in style to the orebody targeted by Western Mining's new search for copper using the altered basalt model. The orebody had no more than 100,000

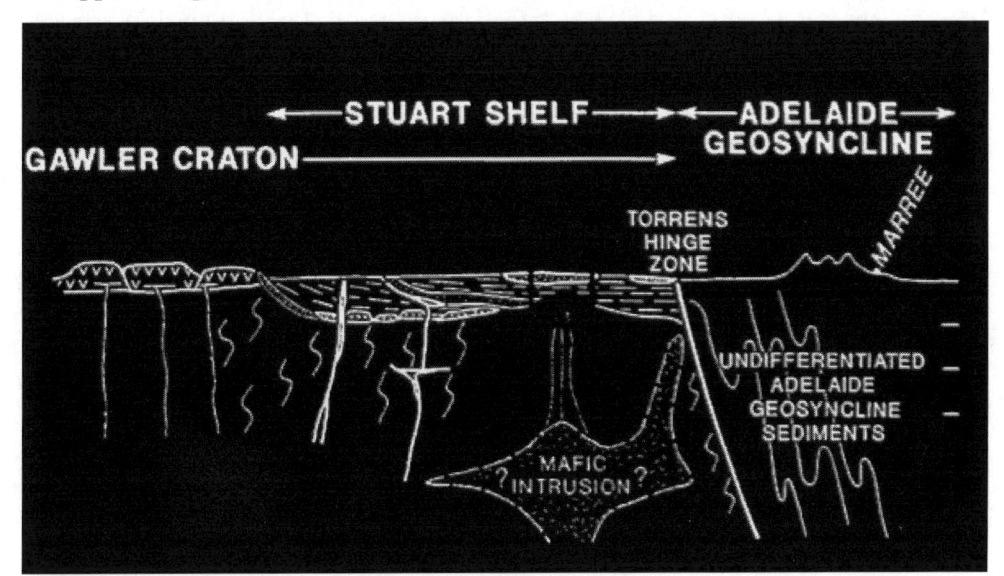

A regional cross-section of South Australia (looking north) from a Western Mining technical handbook in the 1980s, showing the relationship of the Stuart Shelf to the Gawler Craton and the adjacent trough of sediments (the Adelaide Geosyncline) that forms the Mt Lofty and Flinders Ranges.

tonnes of contained copper and was perhaps one percent of the size of the discovery sought by Western Mining, but it was a vital piece of the puzzle. It proved copper minerals were mobile in the Adelaidean sediments and processes had been at work to concentrate these into an orebody.

Geophysical surveys in the Mt Gunson area provided a third vital clue by revealing coincident gravity and magnetic anomalies in the same location as the deposit. The geophysics suggested Mt Gunson sat on top of unusually dense and magnetic rocks, perhaps an uplifted basalt pile of the kind sought by the team. Rutter pointed out that the geophysical anomalies were even larger and more intense in the Andamooka region, which suggested this region held the prospect of a much bigger copper discovery than Mt Gunson. It was a compelling argument and took the search even further north to the Andamooka-Olympic Dam area. Between July and September, 1974, the clues were put together in such a way to take the search almost 350 kilometres north of Roopena to a vast area where deep drilling had never occurred.

Tectonic lineament studies added to the team's confidence in moving so far from conventional copper exploration targets. The lineament studies by O'Driscoll and a structural geologist on his team, David Duncan, had identified a tectonic target at Mt Gunson. This suggested that deep structures existed at this location and had played a part in allowing copper minerals to travel from the deeply buried basalts to the Cattle Grid deposit. O'Driscoll and Duncan had identified many other tectonic targets near Andamooka, adding to the belief this region had all the ingredients for a major copper deposit.

By October 1974, the team finally settled on Andamooka as the first broad target area for exploration drilling. The new search for copper had narrowed from the whole of Australia to an area within a 30 kilometre radius of the opal fields of one of South Australia's most remote towns. The next challenge would be deciding where exactly to hammer a wooden stake that told the drillers where to begin. In a target area that still extended over almost 3,000 square kilometres, there might yet be a thousand ways to let a discovery slip through their fingers.

Late 1973

- Evans agrees to move to Adelaide to establish an exploration base.
- Haynes asks Rutter to provide a regional interpretation of magnetic and gravity data released by the Bureau of Mineral Resources in areas of Proterozoic geology in South Australia, including the eastern margin of Gawler Craton where the Olympic Dam deposit would later be discovered. Rutter is looking for evidence of

buried basalts — either as volcanic piles of basalt or uplifted fault blocks — and faults that might provide a pathway for copper into overlying sediments.

- Lineament tectonic analysis of the northeast Gawler Craton begins.

February 1974

- The Adelaide office opens in the garage of Evans' home at Flagstaff Hill. As officer-in-charge for South Australia, Evans begins many field trips to gain a first-hand understanding of major sequences of rocks that are important under the new exploration model. These are the older Proterozoic basement rocks of the Mt Lofty and Flinders Ranges, the lower Adelaidean rocks in the Roopena-Mt Gunson area, stretching from Whyalla to north of Port Augusta, and the Cambrian sediments of the Kanmantoo Trough, a deep structure that arches around Adelaide from Kangaroo Island to Morgan on the Murray River.
- Haynes and Evans begin inspecting drill core at South Australian Department of Mines, searching for the right types of altered basalt. Samples are sent to Western Mining's laboratory in Ballarat for analysis to identify basalts with the sought-after chemical composition and depletion in copper.
- Rutter completes the first of his regional gravity and magnetic interpretations.
- A budget allocation of about $40,000 in 1974 dollars ($270,000 in today's terms) is made for drilling two stratigraphic exploration holes. "Planned to be used in 2 stratigraphic drill holes on the Stuart Shelf. Siting partially based on geophysical interpretation of basement highs." The provision was deleted a few months later because the definition of drilling targets was still many months away. The provision was made again in late 1974.

April 1974

- Tectonic lineament studies on the Torrens and Andamooka 1:250,000 map sheets are completed.

May 1974

- The exhaustive research of the past year to identify the most prospective targets in South Australia is distilled into a major report, entitled *Stratiform Copper — SA Exploration Strategy*. The report names nine areas scattered across the state — Wooltana in the northern Flinders Ranges, the Willouran Ranges south of Maree, Uro Bluff north of Port Augusta, Tent Hill west and southwest of Port Augusta, Depot Creek northeast of Port Augusta in the Flinders Ranges, a geological structure known as the Torrens Hinge Zone from south of Port Augusta to

Adelaide, Truro northeast of Adelaide, the Eyre Peninsula and the Olary region southwest of Broken Hill.

- Lalor, Evans, Haynes and Rutter make a week-long field trip to Truro, Depot Creek and Roopena, part of the Tent Hill target area west and southwest of Port Augusta. Field inspections reveal these basalts have the right signs of alteration, and samples are taken for chemical analysis.

June 1974

- The Bureau of Mineral Resources releases Bouguer gravity interpretations of regional survey data conducted several years earlier on the Stuart Shelf. The data shows several large and very strong anomalies around Andamooka, an opal mining town on the western shores of Lake Torrens, about 240 kilometres north of Port Augusta. Haynes in Kalgoorlie is the first to see the new BMR maps, and quickly forwards it to Rutter in Melbourne, with a note highlighting the anomalies and asking whether they might fit the new model for copper discoveries. Evans also encourages Rutter to focus on interpretation of the Andamooka anomalies because they have no explanation in the geology of the Adelaidean sediments.

July 1974

- The team is excited by the test results of basalts at Roopena, which show they have lost up to 60% of their original copper content. Analysis of Truro and Depot Creek basalts discourages further search in these locations.

- Rutter sends Haynes his interpretation of the BMR's regional magnetic survey of the Andamooka map sheet, an area of 15,700 square kilometres that includes Olympic Dam. Rutter writes "the magnetic map is considerably more complex than would have been anticipated from the surface geology" and concludes these signals must be coming from basement rocks below the Adelaidean sediments. He breaks the map sheet into 10 zones based on the character of the magnetic readings, and is most excited by Zone IX, which includes Olympic Dam. "This may have been the centre of volcanic activity, with some of the anomalies being the expression of vents filled with magnetic material, and others resulting from flows of magnetic rock. They appear very close to the surface ...".

- Evans makes a five-day field trip to the Woomera area to make on-the-ground inspections of locations where magnetic anomalies and tectonic targets have been identified by mapping. It is the first visit to the Olympic Dam area by a Western Mining geologist in the new search for copper.

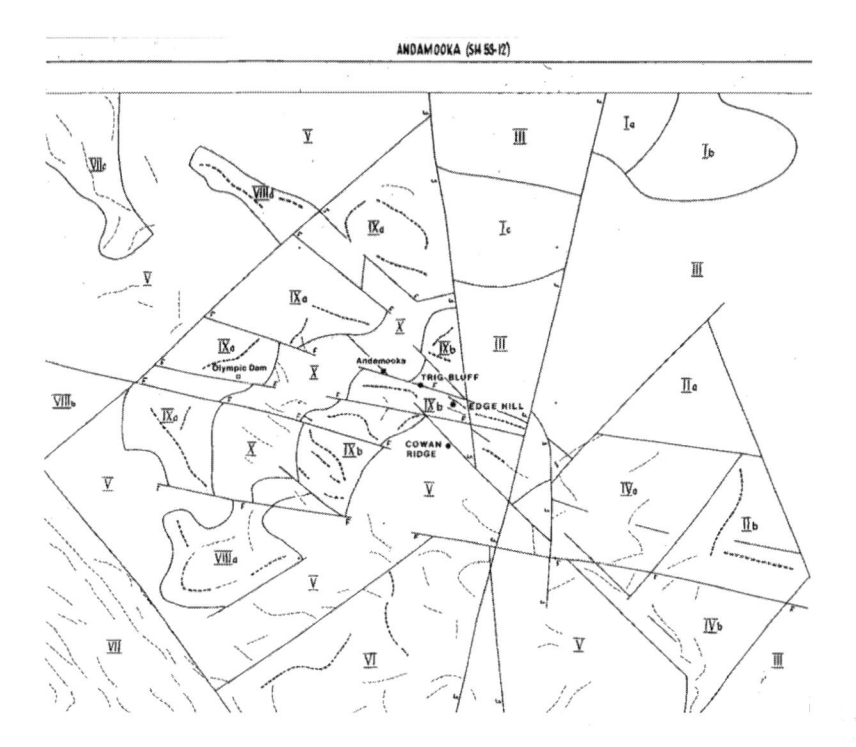

Rutter's interpretation of magnetics on the Andamooka mapsheet, also from the K/2792 report, in which he highlighted Olympic Dam and suggested it might be an ancient volcanic centre.

August 1974

- Rutter consolidates detailed magnetic and gravity interpretations on a map of the Torrens and Andamooka sheets. He notes strong similarities between the gravity and magnetic signatures above the Mt Gunson copper mine and an area about 80 kilometres north, which is almost as far north as Olympic Dam. Tectonic targets are added to the map, which highlights the coincidence of geophysical anomalies and tectonic targets at Mt Gunson.

September 1974

- The work program has narrowed to only two locations on the Stuart Shelf — the Cultana granite area, immediately north of Whyalla, and Andamooka, with the latter clearly causing the greatest excitement among the team. Two other Stuart Shelf locations had recently been rejected. These were Uro Bluff, which showed only weak magnetic and gravity anomalies, and Kingoonya, because it was within the Woomera firing range and access could not be negotiated with the

Government. The latest work program calls for sampling of stream beds at Cultana after an earlier rock chip sample had assayed 1% copper. At Andamooka, plans for are being drawn for an exploration licence application.

- Rutter selects five coincident magnetic and gravity anomalies on the Andamooka map sheet, and a sixth anomaly to the south on the adjacent Torrens map sheet. These targets are selected because Rutter calculates they have shallower source rocks than other coincident anomalies in the region. He is struggling to explain the anomalies in terms of Haynes' altered basalt model, but writes in an internal memo to Haynes that the anomalies are of immense interest and worth drilling no matter what exploration model is being pursued.

- The five targets on the Andamooka map sheet are named after the nearest geographic features. Rutter calculates the estimated depth of the source rocks for the gravity anomaly at each location, based on the difference in density between samples of the Roopena basalts and the overlying Adelaidean sediments. The five targets are Andamooka Island (shown at the location marked "1.7", which indicates the estimated depth to the source of the gravity anomaly in kilometres), the Lake Torrens anomaly ("2.7"), Bill's Lookout ("2.2") within the Andamooka opal field; Olympic Dam ("1.15") and Appendicitis Dam ("4.8"). The sixth anomaly on the Torrens map sheet drops out of consideration. Further calculations suggest the source of the Olympic Dam gravity anomaly could be as shallow as 850 metres, but this is still a very deep and expensive drilling target.

- The lineament analysis begun in March has defined 21 "tectonic targets", including targets over Olympic Dam, Appendicitis Dam and further south at Mt Gunson. The coincidence of gravity, magnetic and tectonic targets at the Mt Gunson copper mine greatly adds to confidence in exploration program near Andamooka, where the geophysical anomalies are even larger.

October 74

- At the Adelaide Base final review on 5 October 74, the team settles on the area around Andamooka as the location for the first exploration licence under the new search for copper. Evans and Lalor are tasked with outlining the proposal to South Australia's director of mines, Bruce Webb.

- The plan is to drill five for six diamond drill holes over two years to test "geophysical and/or tectonic targets". "The probability of success should be optimized by locating diamond drill holes in area where tectonic target coincides with magnetic and gravity anomalies."

November 74

- Evans visits potential drill locations around Andamooka to survey road access and water supplies.

December 74

- Application is made on Christmas Eve for a 2,970 square kilometre exploration licence at Andamooka, with proposed expenditure of $60,000. A revised application is submitted a few weeks later excluding about 600 square kilometres over the Andamooka Precious Stones Field.

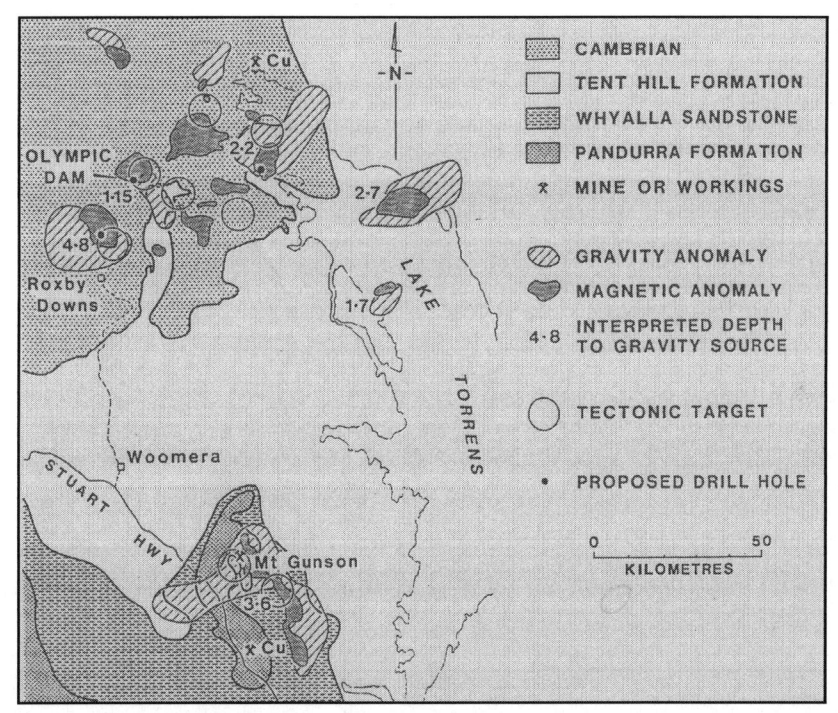

The September 1974 map by Preston office draftsman Wally Churchill that compiled gravity, magnetic and tectonic interpretations near Andamooka for the first time, as presented by Jim Lalor in 1984 in the first public address about the discovery. The map shows three proposed drill hole locations — Bill's Lookout, Olympic Dam and Appendicitis Dam. The map also featured the interpreted location of contacts between two overlying units of sediments — the Pandurra Formation and the Tent Hill Formation. This contact was believed to be a key influence on the location of copper, based on what the team knew from the Mt Gunson mine.

8

Olympic Dam becomes the target

In December 1974, Dan Evans wrote a report seeking the approval of Exploration Division management for drilling in the Andamooka area.

The paper summed up the work of the team over the past 18 months and made the case for the company to risk $25,000 to $30,000 in 1975 dollars on a 700 metre-deep drill hole. In today's terms, this is equivalent to $145,000 to $175,000.

The report is remarkable today because it reminds us how little the team knew about the rocks below the sand dunes and clay pans of the Andamooka region. The team was brought to the area by an innovative PhD, some rocks collected almost 350 kilometres away near Whyalla, geophysical surveys from low-flying aircraft and gravity stations at seven kilometre intervals, and tectonic studies of aerial photos.

Evans had visited the Andamooka region, but it would not have made much difference to his report if he had stayed at home. Field trips told him only about the practical difficulties of drilling in the area, and revealed nothing about the rocks of interest many hundreds of metres below.

The hopes of the team for that first drill hole were clearly conveyed by the title of Evans' paper, *Andamooka Stratigraphic Drilling Proposal,* written in December 1974 and circulated in March 1975. This was to be a stratigraphic hole, the type of drilling more usually undertaken by governments because the risks are too high for private explorers. Governments drill stratigraphic holes to obtain a complete physical log of the rocks from the surface to the bottom of the hole. They are designed to fill gaps in the knowledge of a region's geology and remove some of the risks of exploration. In the best-case scenario, a government-funded stratigraphic hole yields information that encourages private explorers to risk their own money and eventually make mineral discoveries.

In this case, Western Mining would be the pioneer by drilling the first deep mineral exploration hole on the Stuart Shelf — a 48,000 square kilometre area bigger than

Tasmania. The risks were high, but the team knew something about the potential rewards that no other explorer understood. It was prepared to take the gamble.

The *Andamooka Stratigraphic Drilling Proposal* makes it clear the team did not expect to find copper with its first hole. It expected to find the source of the magnetic anomaly, and it hoped this would be basalt. If basalt was present <u>and</u> heavily depleted in copper like the rocks at Roopena, this would qualify as a great success.

The biggest risk with the first hole was uncertainty about the depth at which basalt would be intersected. It was quite possible that Western Mining could bring a diamond drilling rig all the way from Kalgoorlie, drill to a depth of 700 metres and still not intersect anything that would explain the large and intense magnetic anomalies.

The team hoped a drill depth of 600 to 700 metres would be enough, based on an estimate from the Bureau of Mineral Resources that magnetic rocks began at a depth of 500 metres in the Andamooka area. But the report also noted that the basalt target could be as deep as 1,255 metres.

"Due to the flat-lying disposition of the rocks, it is impossible to predict stratigraphic thicknesses without seismic data. Therefore, although the actual units present are reasonably certain, their stratigraphic thicknesses can only be estimated. Preferably (the first exploration hole) would be preceded by a seismic survey which would determine the depth to the magnetic horizon, i.e. basalts?"

The *Andamooka Stratigraphic Drilling Proposal* is also revealing because of its recommendation of a two-phase approach to drilling. It was built around the idea that success would come slowly, and that persistence was an essential ingredient in the ultimate discovery of copper.

Phase I was to be one exploration hole, preferably after a seismic survey to improve confidence in the depth of hole required. If no basalt was found, exploration for copper in this area would end — no ifs, buts or maybes. If basalt was found and the initial drill hole confirmed the presence of overlying shales or fine-grained quartzites, which are suitable hosts for a sedimentary copper deposit, phase II would begin.

Phase II should be a program of five holes, centred on locations of coincident geophysical and tectonic targets. Preferably these should be as deep as the initial exploration hole to provide as much information as possible about the stratigraphy of the region and the geological structures deep below the surface.

Phase II, which was estimated to cost up to $150,000 in 1975 dollars ($865,000 in today's terms), was all about locating the copper deposit that, according to Haynes' model, should be in the vicinity of the altered basalts. All their innovative thinking would count for nothing if they could not track it down from here.

The team clearly knew they faced long odds of making a discovery, even if altered basalts were found with the initial hole. Even with the benefit of geophysical surveys and tectonic lineament analysis, they were dealing with very broad target areas, in the order of 20 square kilometres in size. They could drill 100 holes in ground this size and still miss a major orebody.

To give Western Mining a reasonable chance of discovery, Evans argued in his report for a minimum of five holes, using a theory known as expected value. He learned the theory studying at McGill University under Professor Brian McKenzie, who pioneered many of the key concepts of mineral economics. Evans' argument is fascinating reading today in light of the events that were to follow.

"Expected value is a technique by which the financial benefits of an exploration win can be estimated with respect to the probability of it happening. It is defined as:

$EV = pR - qC$ where p = probability of exploration success

R = net present value given a success

q = probability of failure $(1 - p)$

C = cost of exploration investment

"Overall exploration is regarded as a negatively valued game because the probability of success is so low (normally around 0.001) that $p \times R < q \times C$, making EV negative. Under these conditions, the only justification for participation is believing in the windfall theory, i.e. discovering the elephant quite by accident.

"The equation is most sensitive to 'p' because of its very low value. Increased skills or specialist techniques are basically all aimed at increasing 'p'. Although it is obviously impossible to arrive at a unique solution for the expected value of the proposed exploration programme in the Andamooka region of the Stuart Shelf, it is instructive to solve the equation under a range of conditions."

Evans then calculates a table of expected values. The probability of discovery is varied from 0.001 (i.e. the chance of discovery is one in 1000) to 0.1 (a one in 10 chance). He also varies the value of a potential discovery from a low of $15 million to a maximum of $100 million in 1975 dollars. In today's terms, this is equivalent to a minimum of $85 million and a maximum of $575 million.

History would show that the discovery was made after 10 holes — the one-in-ten chance that represented the very best odds that Evans dared to imagine. Of course, history would also show the program led to a discovery worth very much more than the maximum value of $575 million. As explained in Chapter 2, we know now that Olympic Dam is the largest, single mineral resource in the world, worth more than 1,500 times the very best scenario in the *Andamooka Stratigraphic Drilling Proposal*.

Five becomes two

Western Mining's records of the countdown to drilling in the Andamooka region are extensive. The K/2792 report runs to 164 pages, even after many of the memos and reports of that period were reduced to extracts. But there are significant gaps in the official history that allow controversy to continue over some aspects of the discovery.

One of the gaps is how the team narrowed its target selection from five locations at the end of 1974 to the two prospects that were actually drilled — Olympic Dam and Appendicitis Dam (later known as Acropolis).

In Western Mining's first public address on the discovery, in 1984, Jim Lalor said: "Both of these targets were generally ranked as being the most promising of all the targets, although a detailed written listing of the reasons for such a choice was never made."

The closest we can get to an official account are the reasons documented in the K/2792 report for selecting any of the targets around Andamooka, as Lalor also recounted in his address.

"The reasons for the stratigraphic drilling of these targets was documented as follows:

1. Sediment hosted copper deposit source rocks probably underlay the region, with a local thickening probable near the northern projection of the Torrens Hinge Zone;
2. The Pandurra Formation-Whyalla Sandstone unconformity, on which some copper mineralisation in the Mt Gunson area was located, probably existed in the area;
3. The geophysical interpretation suggested that the targets included anomalies similar, but larger than those that occurred beneath Mt Gunson;
4. Favourable tectonic analysis targets had been defined, four of which coincided with geophysical anomalies."

Roy Woodall says today the role of geophysics in the selection of drilling targets has overshadowed the importance of the O'Driscoll's tectonic lineament studies.

"All of the scientific disciplines we used to locate Olympic Dam were equally important — it was a team effort."

Woodall says that, as the strongest supporter of O'Driscoll's tectonic studies, he was adamant the drilling targets on which he signed off featured a gravity anomaly, a magnetic anomaly and a tectonic target.

He says the decisive work to select Olympic Dam and Appendicitis Dam was completed by the end of 1974, with further field work concerned only with gaining more information about the probable depth to the basement rocks and therefore the type of drilling rig to be brought from Kalgoorlie.

Geophysics and target selection

Hugh Rutter says that of the five targets identified at the end of 1974, his first preference, based on interpretation of the gravity and magnetic data, was Bill's Lookout in the Andamooka Precious Stones Field.

At the time, Western Mining was pushing hard for access to the opal field, where all mining rights belonged to the local prospectors, even though they had no interest in digging below 50 metres. Evans believed the Department of Mines might compromise. He wrote in the *Andamooka Stratigraphic Drilling Proposal* that "most of the region of interest is covered by the Andamooka Precious Stones Field in which mineral exploration is prohibited. However, the field is apparently nearing depletion. Proposals to either alter the boundaries or exempt the upper 100 metres from an exploration license or mining permit might be possible."

Evans added that "a very positive factor to be considered is the attitude of the South Australian Mines Department towards promoting mineral exploration in South Australia. The desire shown by the SADM to encourage and assist is a reflection of the Minister of Mines' and the Premier's commitment to developing the State's natural resources." These words read like a media release, but were contained in an internal memo never intended for the public. The company was genuinely impressed and thankful for the assistance it had received in its new search for copper in South Australia.

The opal prospectors of Andamooka certainly did not share the view the field was nearly depleted, and the State Government was in no mood to pick a fight with them while Western Mining had other targets it could pursue. The company submitted on 11 February 1975 an amended application for an exploration licence that excluded the Precious Stones Field. The application provides some evidence of how targets were being narrowed down by stating that "theoretical geophysical calculations will be made to select the shallowest coincident gravity/magnetic anomaly outside the Precious Stones Field."

Rutter says the focus of the search shifted to the next best target outside the opal field. "I came up with a second target, which was Olympic Dam. It had the right magnetic signature and gravity signature. I also pointed at this stage to Appendicitis Well, which was renamed Acropolis, and to targets at Arcoona and place called Bopeechee, which is another stock watering well."

He says he wanted better data than the BMR had provided so he could be more confident about the depth of the source of the magnetic and gravity anomalies. It was hoped that altered basalts buried beneath the Adelaidean sediments of the Stuart Shelf were the source of anomalies, but were they too deep to reach by exploration drilling?

If these basalts were deeper than the 600 to 700 metre range of diamond drilling rig, Western Mining would need to rethink its exploration plans.

Rutter says the available data "was atrocious by today's standards. The magnetic data was based on lines at one kilometre intervals, taken from 300 metres in the air. The gravity was based on a seven-by-seven kilometre grid. That's one reading every 50 square kilometres."

He approached the senior geophysicist at South Australia's Department of Mines, Bernie Milton, in August 1974 to discover whether the department held geophysical data that might assist Western Mining. The department had conducted a seismic survey in the region in 1969 for a company by the name of Stuart's Bluff Minerals.

"Bernie suggested we try some seismic and I said it was too expensive. He said: 'I will get our guys out there. They have been sitting around doing nothing. It will do them some good to get out there and do some field work.'"

Seismic surveys are used in the oil and gas industry to generate a subsurface picture of rock layers and the location of potential traps for hydrocarbons. Sensitive instruments

South Australia's pioneering geophysical surveys

The regional geophysical surveys of the Andamooka and Torrens areas might be primitive by today's standards, but Western Mining was lucky to have any data of this kind in the early 1970s. South Australia was at least 10 years ahead of other Australian states in systematically conducting geophysical surveys, thanks largely to Sir Thomas Playford, the State Premier from 1938 to 1965. Sir Tom was zealous about developing South Australia's energy and mineral resources and had a deep interest in exploration. He would often spend his summer holidays with geoscientists on field trips in far north of the state. The Department of Mines, under its newly appointed director, Ben Dickinson (later Sir Ben), set up a dedicated geophysics department in 1948. By comparison, Queensland did not establish a geophysics branch until 1959, while New South Wales and Tasmania would wait until 1961 and 1963 respectively. In its first five years, South Australia's geophysics unit flew five major surveys covering iron ore deposits west of Whyalla and the copper mines of Moonta. The newly formed Bureau of Mineral Resources flew the surveys for the department with its specially equipped DC-3 aircraft. The aeromagnetic survey of the Andamooka and Torrens area, conducted in 1961-62, was the eleventh major survey of its kind in South Australia, according to former director of the department, Keith Johns. A regional gravity survey followed in 1969, just four years before Western Mining would turn its attention to the region.

record the reflections and refractions of high-energy sound waves. The time taken by sound waves to return to the surface and their angle of reflection creates a surprisingly complete picture of buried rock strata. Seismic is a proven exploration method in areas where oil and gas are typically discovered, but these rocks are geologically young at about 250 million years of age and not subjected to high pressures and temperatures.

In the much older and compacted Adelaidean sediments of the Stuart Shelf, the effectiveness of a seismic survey was an open question, but Milton told Rutter it was certainly worth a try. The Adelaidean sediments had sat remarkably undisturbed on the ancient and stable chunk of continent known as the Gawler Craton for up to 1,000 million years. If it did work, a seismic survey might provide an accurate picture of the depth of the boundary between any buried basalts and the overlying sediments.

Milton's team for the Olympic Dam survey included Reg Nelson, who joined the department a couple of years earlier after completing a geophysics degree at the University of Adelaide. Nelson today is one of Australia's most successful exploration geophysicists and managing director of Beach Energy. "I had become involved in developing high resolution seismic reflection techniques, which led to my involvement in running these over Olympic Dam and later over other areas of the Stuart Shelf, as well as Mt Isa in Queensland." He says Milton was a driving force in assembling state-wide maps of gravity and magnetics and he did much to advance the understanding of South Australia's geology, including the stratigraphy and structure of the Adelaidean sediments.

Western Mining's records show Milton offered to conduct a limited seismic survey for the company over four days at a mutually agreed target near Andamooka. "As the work would be of an experimental nature all costs would be met by SADME." It was a very convenient outcome for Western Mining's lightly resourced copper exploration project. Milton's offer was later watered down to the sharing of costs, but there was still a substantial financial contribution by South Australian taxpayers.

In April 1975, just two weeks before the survey was scheduled to begin, the director of mines, Bruce Webb, sent a written quote to Western Mining for a four-day experimental survey, followed by a six-day production survey. Webb wrote: "Because of the experimental nature of the investigations and the possible wide ranging implications if useful results are obtained, it is necessary to operate with a crew larger than would be normal for production work and containing senior personnel, and using equipment which has been designed primarily for sedimentary studies in petroleum exploration. As a result, the quotation for this work does not contain salaries and wages of the crew of five, and equipment hire is not charged for the first four days operation."

The quote totalled $5,110 or about $30,000 in today's terms, and comprised of $1,420 for the four-day experimental survey, $2,700 for a six-day production survey and $990 for mobilisation charges. The Department of Mines quote represented a major bargain, based on private sector costs at the time for comparable surveys. Evans had included in the *Andamooka Stratigraphic Drilling Proposal* an estimate of $25,000 in 1975 dollars for a seismic program of similar size. And in 1976, Western Mining paid a private sector company $20,000 for a four-day reflection survey elsewhere on Roxby Downs Station.

Based on these comparisons, the Department's subsidy was worth much more than the waiver of some labour and equipment hire. In fact, the contribution by South Australian taxpayers was probably between $80,000 and $120,000 in today's dollars.

Keith Johns, who was then deputy director of mines and deputy government geologist, says today it was not unusual for the department to subsidise private exploration. "Our assistance to Western Mining wasn't anything special. For example, we had contributed to exploration by Santos around Port Augusta and later in the Cooper Basin. It certainly couldn't be assumed we would assist, but if a particular project could benefit other explorers or had some special significance, we were open to the idea. In most cases we could give advice and the benefit of our knowledge about a particular area, but sometimes it was financial support as well. I clearly remember we had doubts

> **Keeping up the PACE**
>
> The modern-day Department of Mines is known as the Minerals Division of the Department of Primary Industry and Resources South Australia, or PIRSA. It continues the Department of Mines tradition of kick-starting mineral exploration in the state by subsidising private explorers and making South Australia an easier and more attractive place to search for orebodies. PIRSA and the South Australian Government take the enlightened view that encouraging the creation of wealth through mineral exploration is a smart strategy in a dry state with limited arable land and a shrinking manufacturing base. In 2004, PIRSA introduced a Plan for Accelerating Exploration (PACE) that has since been copied in other states and even adopted overseas. One of the guiding principles of PACE is the state needs to work harder to attract mineral explorers because of the blanket of barren sediments over much of the prospective areas, particularly the Gawler Craton. It's exactly the same thinking that led department of mines director Bruce Webb in 1975 to authorise a generous subsidy for experimental seismic over Olympic Dam. In its first five years, PACE provided $10 million in subsidies for exploration drilling at 168 projects, including the 2005 discovery of the massive Carrapateena copper-gold deposit not far from Olympic Dam. PACE helped to double the state's share of national spending on mineral exploration to 12% by 2008.

about whether seismic would work on the Stuart Shelf, so we had quite an interest in establishing whether this exploration method could be used in that part of the state."

If any South Australians had been unhappy about taxpayer funding of private mineral exploration, they could not have complained for long; nearly all of the money was spent right over the newly defined Olympic Dam prospect, including the only reflection seismic survey in the 10-day program. The results would prove to be more important than anyone could have predicted. Rutter says: "In my mind I had really decided I wanted to put a drill hole somewhere around Olympic Dam, it was just whereabouts. I really couldn't pick it to the square kilometre at that stage. It was vague, I didn't have much data and I was hoping the DME seismic would improve things. And it did."

The Department of Mines survey was planned in October 1974, but did not get into the field until 9 May 1975 because of unusually wet weather in late 1974. Rutter and Western Mining's senior geophysics field technician for Eastern Australia, Terry Brooks, joined Milton and the Department of Mines crew in Port Augusta the previous evening after a long day's drive from Melbourne.

Rutter and Brooks made the journey in an orange-painted ute that was standard equipment for Western Mining field work. Before leaving the Preston office, Brooks loaded the vehicle with geophysical equipment that he and Rutter would use to conduct their own, more modest surveys while the Department conducted its experimental seismic elsewhere on Roxby Downs Station. The pair planned to take detailed magnetic readings on the ground in the hope this might yield better data for calculating the depth of the source of the magnetic anomaly than the BMR's readings from an aerial survey.

"The idea was to start at a position south of Olympic Dam, and just using a compass and magnetometer walk up to Olympic Dam and beyond to Coronation Dam. We were also going to do a western traverse, which went from east of Olympic Dam and through it to Appendicitis Dam. I had these marked on a plan. We had no GPS in those days, just a map and compass to find our way around."

Rutter says the Department of Mines convoy was bigger than any geophysical survey crew he had seen, with six Land Rovers, three caravans, and a vacuum drilling rig for sinking shot holes in which the explosive charges were laid. "It was a big event. They even had their own cook." The SADME set up camp on 11 May at Burgoyne Ridge, about 12 kilometres south of Olympic Dam.

When the operator of the drill rig became ill, Rutter temporarily took over the role. "I was the only one left who knew how to operate the vacuum drill, and there was no way I was going to allow everyone to turn around and go back to Adelaide."

Rutter and Brooks set up their own campsite near Roxby Downs Homestead, another 12 kilometres further south, and called in to visit station owner, Tom Allison. Evans had visited earlier in the year to advise plans for the geophysical surveys. He had also arranged Commonwealth Government passes for Rutter and Brooks to enter Roxby Downs Station, which was within the Woomera Prohibited Area.

At dawn on 12 May, Rutter and Brooks started the 30-kilometre drive from their camping spot to Olympic Dam, close to where they planned to begin their magnetic ground traverse. Rutter was behind the wheel, which Brooks would soon regret.

"It was the crack of dawn as I was going over this last little ridge just 20 metres from Olympic Dam. I had the sun in my eyes and I went too fast over some rocks, which we hit with a great thud. I drifted down and stopped near the fence at the dam. We were there, but what had happened to our vehicle?

"Terry checked under the car and could see a hole in the sump. It was only a small one, but oil was dripping all over the place. And that wasn't the end of it. Before he could even tell me the bad news, Terry came out shrieking from under the car. I had stopped the car on an anthill and he was covered in ants as well. He wasn't happy with me!"

Brooks knew enough about the vehicle to realise a simple repair was not possible. The engine had to come out before the sump could be replaced.

"We had organised at the end of our walk to meet up with a field assistant at Coronation Dam, so we thought rather than walk back to our camp it would be quicker to carry on and do our profile of magnetics up to Coronation Dam. So we set off. I don't think we were speaking at that stage. This was me who had stuffed up!"

The two men would stop every 25 metres to take a reading with a magnetometer, a small instrument fixed to the top of a pole about 1.5 metres long. Each reading takes only a minute, and the distance between measurement points is simply paced out.

"We were both pacing. It was about 25 paces for Terry and about 30 paces for me, allowing for a few slips on the sand dunes as well. Terry was checking this as we went."

This made it possible for the men to complete a 16-kilometre traverse in a single day. They would actually walk more than 20 kilometres because the starting point of the traverse was four kilometres southeast of their broken-down vehicle.

While the survey method might seem rudimentary, it was certainly not low-tech. Hand-held magnetometers are highly sensitive. Any metal items carried by Rutter or Brooks would interfere with the readings, so they had to wear pants without belts and leave behind their wristwatches. Even the metal eyelets in lace-up shoes could cause problems, so elastic-sided boots were the footwear of choice. A handheld magnetometer even detects a field from the person taking the reading, which means the operator should always stand in the same position relative to the instrument.

Rutter says the day's traverse had some memorable moments. The first of these was the discovery of an ancient tobacco tin under a tree where they paused for a rest. "Someone had been there before us in this remote place. We realised that anyone around there would obviously be attracted to that tree because it was the only shade."

The next surprise was just as flukey, but more dangerous. Tom Allison had warned them the day before about a bad-tempered camel in the area where they were headed. "Tom said: 'Fine, off you go, but beware, there's a wild bull camel out there. Don't go near it!' We thought there was no way we were going to come across one camel out there in that vast expanse!' Well, we had been walking over sand dune after sand dune. We just got over one ridge, and part of the sand dune seemed to rear up and growl at us. Here was this nasty bull camel. He had been fast asleep and now its nose was just three feet away from my face. I didn't know what to do with it. We both sort of eyed each other and thank goodness the camel just snorted and walked off in the other

Stuart's race against Burke & Wills

The old tobacco tin discovered by Rutter and Brooks might have belonged to John Macdouall Stuart, who passed through the Olympic Dam area many times on his way to becoming Australia's most accomplished explorer. On his first expedition from Adelaide in 1858, the Scottish-born civil engineer and mineral prospector discovered freshwater springs 80 kilometres north of Olympic Dam, at a place he named Chamber's Creek (later renamed Stuart Creek). He returned a year later and surveyed the area as part of a deal with the Government to obtain a pastoral lease. Chamber's Creek became an inland staging post for a series of expeditions by Stuart that would culminate in the first return crossing of Australia from south to north in 1862. Stuart was in a race to reach the top of Australia against a much better resourced expedition from Melbourne, led by Robert O'Hara Burke and William John Wills. Governments in South Australia and Victoria sponsored the expeditions because each wanted their capital city to be the destination of a proposed telegraph line from Darwin. Stuart learned of the death of Burke and Wills on returning from his fifth expedition in September 1861. He immediately began planning his sixth and final expedition, which succeeded in reaching Chamber's Bay, east of where Darwin would later be located, in July 1862. He returned to Adelaide in December 1862 without the loss of any of his party of 10 men. The expeditions took a heavy toll on Stuart's health, who died in 1866, aged only 50 years. Stuart's remarkable skills as an explorer allowed Adelaide to win the overland telegraph and cleared the way for roads and railway across the heart of Australia from Port Augusta to Darwin.

direction. Good grief! The chances of us meeting this thing!"

The third surprise during the traverse was highly relevant to the search. Rutter found a fragment of sandstone with a tiny speck of malachite, a distinctive green mineral of copper carbonate. The copper minerals that Western Mining would eventually find deep below Roxby Downs belonged to an entirely different family of copper minerals — copper sulphides that were either black and lustrous or glinted like pyrite in the light.

"I don't know how it got there, but it persuaded all of us, including Jim Lalor, that there was copper around and we were looking in the right place. None of us, in hindsight, know where this piece came from. It was almost as if there was a God up there, and he's put this piece on the ground to keep us going."

The day finished much better than it started, with Brooks discovering that Rutter — the Englishman who always copped a ribbing for his lack of bush skills — had navigated a direct line to Coronation Dam.

"We had it worked out that if we missed Coronation Dam, we would come across a boundary fence. And if we hit that boundary fence, we would go west, then south, then meet the road that would take us back to Olympic Dam. So we weren't going to get lost. We had a fail-safe plan, but lo and behold, here's me sighting through the compass to make sure we are right, and what do we see but a windmill going round. We had made it almost directly. Terry couldn't believe it was right in front of us."

The Western Mining field assistant was waiting there as planned. They drove back along the access track to Coronation Dam, which looped around to join the track to Olympic Dam and their orange ute. The stranded vehicle was towed 30 kilometres back to the campsite just south of Roxby Downs homestead.

With the help of the field assistant, Rutter and Brooks were able the next day to complete their second magnetic traverse from east of Olympic Dam, through Olympic Dam and then southwest about 20 kilometres to Appendicitis Dam.

The next challenge was to get home with the damaged ute. "All we had was a piece of rope and we didn't fancy using that to tow the vehicle all the way to Port Augusta for repairs. It could concertina and crash the cars together. That's when Terry said: 'Oh well, I will use some chewing gum.' He used to chew gum, which was just as well. 'I will block the hole in the sump with chewing gum, fill it up with oil, and get down to Port Augusta.'

"He got it down to Port Augusta all right, and then he thought: 'Bugger it, I'll go to Melbourne'. He ended back in Preston where he had it repaired. He went through about six packs of gum and 20 litres of oil, but he made it."

Rutter flew from Port Augusta back to Melbourne. Brooks would return to Olympic Dam within a matter of days with a land surveyor to take gravity measurements at 100 metre intervals along most of the east-west traverse and part of the north-south traverse. Each measurement point — known as a gravity station — had to be accurately surveyed for its height above sea level.

The gravity and magnetic data from Western Mining's own surveys was back in Melbourne by late May, but it still did not provide the answers Rutter wanted. "I was still getting ambiguous answers. I didn't like to tell everyone, but I could get depths that went between 1.2 kilometres to 2.1 kilometres. I thought no-one is going to like this, who is going to drill down a kilometre? And because there was that range, I was unhappy."

Rutter's recorded his concerns with some understatement in a memo to Evans on 22 May 1975. He wrote that measurement of the magnetic anomalies at Olympic Dam and Appendicitis Dam from the ground traverse showed very similar results to the airborne measurements by the BMR, which suggested a source far deeper than the rocks they could test with drilling. There was more bad news. The shape of the magnetic anomaly was too smooth to be explained by an abrupt end to a sheet of basalt, their desired target under the new model for copper discovery. Rutter concluded: "A lot depends on the seismic results."

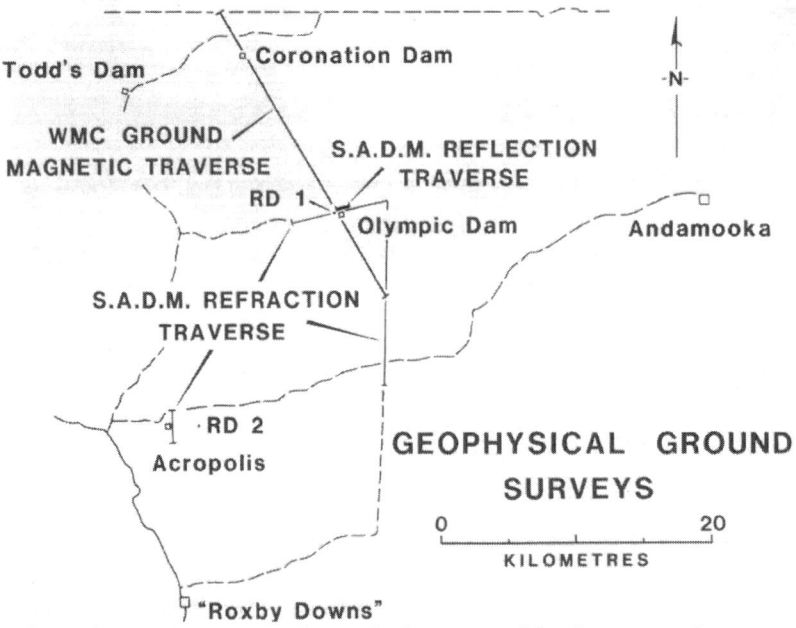

This map from the K/2792 report shows the location of the first ground surveys at Olympic Dam in May 1975. SADME conducted three seismic refraction traverses over a total distance of 20 kilometres and one seismic reflection traverse directly over Olympic Dam. The first ground magnetic traverse up to Coronation Dam by Rutter and Brooks is also shown.

Two weeks later, on 2 June 1975, Rutter flew to Adelaide to look at the preliminary results of the SADME seismic survey with Milton. Rutter wrote the reflection seismic "suggested the presence of a significant event at 300 to 400 metres depth. Hopefully, this will coincide with the top of the magnetic horizon." The existence of the boundary was difficult to interpret from the field results, but Milton was confident of its presence over the full one-kilometre length of the reflection survey over Olympic Dam.

It seemed Rutter finally had the evidence he needed to back a confident recommendation to drill the Olympic Dam prospect. From his interpretation of BMR data and his own magnetic and gravity surveys, he knew the anomalies at Olympic Dam had the shallowest source relative to all of the other prospects being considered. With the seismic data now in hand, he had for the first time what he believed to be a reliable measure of the absolute depth of the source at Olympic Dam. An exploration hole here represented the best chance of determining whether Haynes' new model for discovering copper could bring rewards in the sand dunes west of Andamooka.

Over the next few days in the Preston office, planning for the first exploration holes reached its final stages. A budget allocation for two exploration drill holes had been re-instated about three months earlier, although Rutter and Evans were not aware of this at the time. Rutter says he and Evans had rehearsed a presentation and were unsure about whether they would get any funding. "Evans gave a talk on the geological setting, the model we were looking for, and how Douglas Haynes had said this a good area, we've got the right type of Roopena rocks and I'll pass over to Hugh.

"I jumped up and said we've done the geophysics, we've got the seismic and looked at various places. Olympic Dam we think is the first place, Appendicitis was next, and the first target is at a depth of about 335 metres.

"I sat down, Dan got up and summarised it. Well, Roy really played it. He looked at us and said: 'I'm not going to allocate funds for one drill there.' He paused. 'I'll give you two!' Our hearts sunk then rose. And he just laughed, He knew he was having us on. I remember that day."

The drilling campaign was set to go. The only remaining question was the depth of the first drill hole. Should it stop as soon as hoped-for basalts were intersected and thereby save precious financial resources, or should it continue for up to 150 metres into the basalt to provide generous samples for testing of copper depletion?

A compromise was reached — the drilling would continue for 50 metres beyond the basalt intersection. At the time, no-one could know this decision would mean the difference between disappointment and the discovery of the world's largest, single mineral resource.

⚒ 9 ⚒

Discovery — at last

The 1,100-kilometre drive across the desolate Nullarbor Plain is testing in the best of circumstances. The hardships are even greater in a 1950s-vintage Bedford truck with a makeshift drilling mast on top. But there were compensations for Western Mining driller, Ted Whenan, and his wife, Shirley, on the long drive in their mobile drill rig over the Nullarbor from Kalgoorlie to Roxby Downs in June 1975. At least it was sealed road for all but the last 70 kilometres of the 2,100-kilometre journey, unlike the dusty roads and near-unrecognisable tracks they usually travelled to drilling sites outside Kalgoorlie — their base for the past seven years.

Ted could also take the easier option of leaving his drilling helper to drive the Bedford and join Shirley in the Toyota traytop that towed their 15-foot Roper caravan wherever they went. Perhaps the cheeriest thought of all for the Whenans during that long drive was the idea of heading home to South Australia. The couple were descendants of Cornish miners who flocked to the Moonta region in the 1850s when it was the largest supplier of copper to Europe. After working most of his life as a farmhand, Ted stumbled on a new career as a driller when Western Mining and its partners came to Moonta in 1960 with millions to spend on the search for new riches of copper.

Ted learned quickly and over the next decade established himself as one of the company's most talented diamond drill operators. Western Mining quit Moonta in 1970 after finding only small ore bodies, but the project did make two important discoveries. One of them was Ted Whenan, the other was the introduction of low-cost, mobile drill rigs just like the one now being brought to Roxby Downs.

On 10 June 1975, the Whenans and their assistant approached the final gate on their long journey to Olympic Dam, one of 23 man-made watering holes on Roxby Downs Station. The previous pastoralist on the station, Dave Greenfield, dug the dam out of a claypan in 1956. It was the year of the Olympic Games in Melbourne and clearly Olympic fever had reached even this remote part of the world.

As they approached, the Whenans saw a small sedan stopped at the gate and its driver waiting by the vehicle. His choice of transport was more suited to a shopping centre car park than the inhospitable dirt tracks all the way out here. But the driver was not lost — it was Hugh Rutter, the Western Mining geophysicist they knew from nickel exploration drilling a few years earlier at Kambalda. Rutter explained he was on his way back from pegging the site where the Whenans were to drill Roxby Diamond 1 or RD1. He had driven up from Port Augusta that morning, after flying the previous day from Melbourne to Adelaide and driving 320 kilometres to Port Augusta. The rushed trip followed an urgent phone call from Dan Evans, the South Australia exploration chief.

Evans would normally take care of pegging drill sites in South Australia, but says he had recently received rush orders of his own. He was already in the field leading a secretive "exploration blitz" in the remote Peake-Denison Range, about 150 kilometres southeast of Oodnadatta. This followed the announcement of a massive gold discovery at Telfer in Western Australia by the US gold mining company, Newmont. "Western Mining had explored Telfer for copper a year earlier, but had made a major error by failing to test for gold in any of the dozens of rock samples collected.

Ted and Shirley Whenan at an Olympic Dam reunion in 1995. Photo by Dan Evans

"Roy Woodall dragged most of his senior geologists from all over the country to Kalgoorlie to suggest Telfer-type exploration targets. I proposed a program over the Peake-Denison Range area and within days was in the area with other company geologists. The three-week program was in mid-stream when the Whenans were due to arrive at Roxby Downs."

After his chance meeting with the Whenans, Rutter turned around his car to escort them to the peg he had hammered into the ground at 1 p.m., less than half an hour earlier. He had flown from Melbourne with metre-long wooden stakes in his carry-on luggage (one to mark RD1 at Olympic Dam, and other to mark RD2 at Appendicitis Dam, about 20 kilometres southwest). The stakes were topped with white paint and a small yellow flag, but Rutter was still worried his pegged locations might be easy to miss.

He had picked a site 150 metres northwest of the Olympic Dam livestock watering hole. Rutter had known for months he wanted the first exploration hole to be around Olympic Dam. In the end, the precise location was a compromise. "I would have liked to put the peg right by the dam because that was where we had measured the most abrupt change in gravity readings. With our exploration model, the best copper exploration target was on the steepest part of the gravity gradient. But Tom Allison had told me earlier that day — I had collected his mail in Woomera and dropped it off on the way through — to make sure the drill was not so close it kept his livestock from coming in for a drink. When I asked him how far that needed to be, he told me 150 metres. So that's exactly how far I moved away. I went northwest from the dam because it was on the trend I wanted and it was a flatter area, out of the sand dunes."

The dam was much more than a landmark; it was a vitally important supply of water for the thirsty drilling rig. Diamond core drilling consumes about 10,000 litres of water every day. Water is pumped down the hollow centre of the drill stem to cool and lubricate the diamond-studded bit as it grinds into the Earth. The water then returns to the surface, flushing out large amounts of grit and rock fragments. While a diamond drill extracts a neat core of rock from the entire length of the hole, it still grinds up roughly the same volume of rock around the core. Water pumped out of the hole is piped to a "sump" or shallow pit dug by the driller next to the rig. The grit and silt settles to the bottom. Clear water that does not soak into the ground or evaporate can be re-used. The sump is back-filled when drilling is finished.

In arid areas such as Roxby Downs station, it would not have been unusual to bring in a water tanker to meet the driller's needs. But the first exploration drilling in Western Mining's new search for copper was being done on a tight budget. Drilling targets were not selected because they were the best geologically, but because they were geologically favourable *and* accessible by existing tracks *and* adjacent to a water supply.

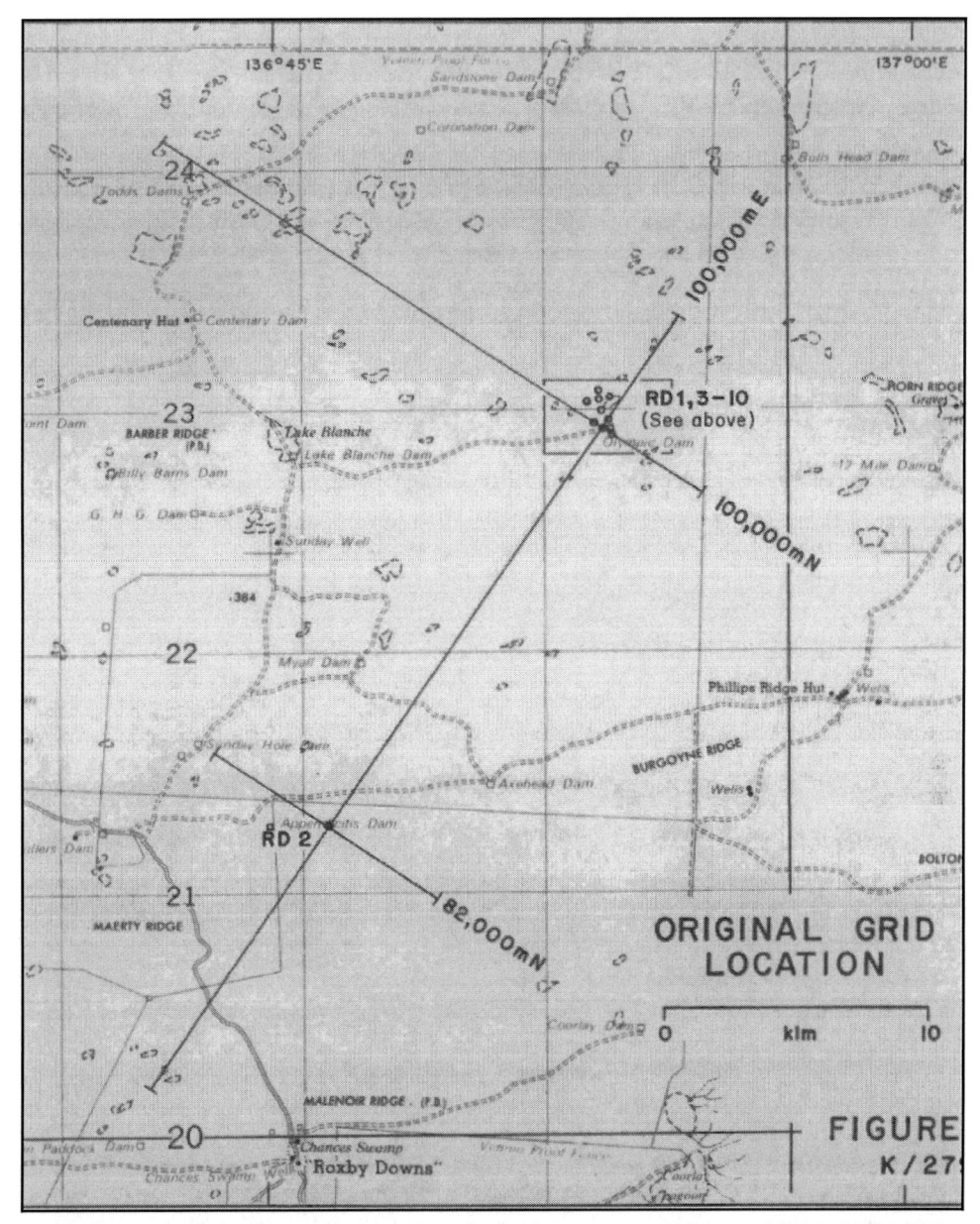

A regional-sized grid was drawn up to fit the locations of RD1 and RD2. When RD2 at Appendicitis Dam failed to reach the basement, the search focused on Olympic Dam on a reduced grid. Source: BHP Billiton archive.

Tom Allison had agreed to a request from Evans to allow Western Mining to draw water from Olympic Dam for RD1 and from Appendicitis Dam for RD2. Allison had good water stocks at the time because it was the middle of winter and South Australia's far north had experienced unusually heavy rains in the past two years.

Rutter was keen to move on after showing the RD1 site to the drill crew. He still had to select and peg the location for RD2, which might take until mid-afternoon, before starting a six or seven hour drive back to Adelaide in his unlikely little sedan.

Ted Whenan and his offsider began digging the sump and laying 150 metres of one-inch diameter pipe to bring water from the dam. This pipe would become the basis of one of the myths around the Olympic Dam discovery. According to folklore, Whenan did not have enough pipe so he moved the rig much closer to the dam. The inference of the story is the discovery of the world's largest, single mineral resource was made because Whenan did not have a few more dollar's worth of plastic pipe. But drilling supervisor, John Emerson, says the story is untrue. He confirms RD1 was drilled within a few metres of the pegged location. He also dispels another myth that RD1 was drilled somewhere other than the intended location because Whenan's truck became bogged.

Western Mining's forgotten exploration heroes

Western Mining made a breakthrough in drilling costs at Moonta by converting retired Bedford buses into mobile drill rigs. By incorporating a number of recent innovations in drilling technology into the converted passenger buses, drilling costs were slashed from $7.50 to $2.50/ft (in 1967 dollars). This effectively tripled the number of exploration holes Western Mining could drill without a budget increase. The home-made drilling technology was taken to Kalgoorlie, but based on old Bedford trucks that could better negotiate the local terrain. It's not known whether any of these rigs were preserved, but they earned a place in mining's hall of fame. Western Mining geologist Jim Lissiman, who took over as manager of the Moonta project in 1964, writes about the rigs in his memoir, *Lucky Jim*. "I looked at the infrastructure that we had at Moonta and felt that we could design and build a rig there to accommodate all these new ideas. We had an old mechanic on staff there, Norm Draper, who had been the workshop foreman at Champions Ltd, a big motor dealer in Adelaide ... He could turn his hand to anything. There was also a small metal fabricating shop at Moonta run by George Thatcher who was the best welder I ever saw. Most of the time he was building TV antenna towers which were essential around Moonta because of the distance of the transmitters in the Adelaide hills, and building a drilling mast and the other fabricating work needed was right up his alley."

After a few days of preparations, drilling got underway on Saturday 14 June, 1975. The rig ran on a single-shift basis, beginning at 6 a.m. and finished at 4 p.m., six days a week. Drilling progressed at a typical pace of about 10 metres a day. If the recent seismic surveys of the South Australian Department of Mines could be relied on, Whenan could expect the first 30 or so days to be uneventful. The survey suggested flat-lying, Adelaidean sediments down to 330 metres, then a major change of rock type to something unknown. Western Mining's geoscientists hoped the rocks below 330 metres would be dense, altered basalts, but so little was known about the deep rocks in this area that anything could show up.

Once or twice a day, drilling would stop and a cylindrical core of rock would be pulled to the surface by hoisting a metal sleeve inside the drill stem. The driller's helper would break the core into lengths of about a metre and place the pieces in steel trays of U-shaped grooves. He would carefully log the below-ground depth of every section for later inspection by the company's geologists. Unlike some of the big exploration companies, Western Mining could not afford to have a geologist sitting on the rig throughout the drilling time. Whenan was in two-way radio contact with Emerson and would make daily reports on the type of rocks being drilled. Emerson would also make two or three trips to the rig, sometimes accompanying the deliveries of more lengths of drill pipe and diesel fuel for the powerful motor that spun the drill bit.

About a month after drilling began, RD1 intersected a major change in rock type at 335 metres — almost exactly where the South Australian Department of Mines had predicted. The monotonous limestones, sandstones and shales of the Adelaidean group suddenly gave way to a dark-coloured rock. It looked like they had hit the geological basement of the ancient Gawler Craton, but what rocks were these?

Drilling continued for another 76 metres until RD1 finished at a depth of 411 metres on 30 July, 1975. The Whenans immediately moved the rig to Appendicitis Dam to drill RD2, but this time they did not break through the Adelaidean sediments. RD2 continued to 513 metres, more than 100 metres deeper than RD1. Later drilling would show the basement was another 47 metres below — perhaps only five more days of drilling. However, Ted Whenan had already pushed the Longyear 38 drill rig to its limits, and well beyond the depth a less-skilled driller could achieve. For now, the last 76 metres of RD1 would be the only basement rocks they had to examine.

The drill core from RD1 was trucked 580 kilometres to Dan Evans' backyard at Flagstaff Hill, Adelaide. He could see immediately there were no visible copper minerals. While the team had always maintained it could take many drill holes to find copper, in their hearts they secretly hoped for some sign of copper from RD1 or RD2. But any disappointment was tempered by what looked like altered basalts at the bottom of RD1, which was the next-best result and a tremendously encouraging start to their radical new search for copper.

Douglas Haynes says he clearly remembers the events that immediately followed, and has diary notes to fill out some of the details. "When the drill core arrived at Flagstaff Hill, I was in the field in northwest WA. Dan contacted me by phone when I went into town, probably Paraburdoo, and Dan's first words were: 'Doug, we have the most fantastic altered basalts in the bottom of Drill Hole RD1 — what do we do now?' I indicated that I would come straight over to the Flagstaff Hill office when I finished the current spell of field work. I was based in Kalgoorlie then."

The only known photo of Ted Whenan's mobile drill rig at Roxby Downs is this shot taken by Hugh Rutter while the rig was drilling RD3 in October/November 1975. The rig drilled only one more exploration hole (RD4 in December 75 and January 76) before Whenan was given a more powerful Longyear 44 rig purchased from Broken Hill South. Rutter was back on site when drilling of RD3 began to conduct more magnetic and gravity ground traverses with Terry Brooks.

DISCOVERY AT LAST

In the meantime, Evans conducted geological logging of the core, which means measuring, describing and where possible identifying the rock units from the top to the bottom of the hole. Evans could correlate most of the major rock units with limestones, sandstones and shales of the Adelaidean sediments that were very familiar from his many field trips in South Australia. Evans also took a chip sample every 15 centimetres along the RD1 core. The samples over every two-metre interval were bagged together and sent to Western Mining's geochemical laboratory in Ballarat for testing for the all-important traces of copper.

Haynes' fieldwork prevented him from making the trip to Adelaide until the third week of September, almost two months after the core from RD1 arrived at Flagstaff Hill. By then the core from RD2 had also arrived in Evans' backyard. "In the cold, Dan and I went out in the backyard to look at the core. Dan did not know for sure what the rocks at the bottom of RD1 were — and nor did I — we both thought they were some type of complex, broken-up volcanic rock. It was cold, and raining, and we returned to the office to have a cup of coffee. At that exact time, the handwritten assay results arrived by post from the assay laboratory in the Western Mining mailbag. Dan looked at the results, and said: 'Look at this, look at this'. We were completely taken by surprise, and need I say, exhilaration."

The assay revealed a 38-metre intersection averaging 1.05% copper. It was not an economic grade of copper, but an astonishing result from the first drill hole in a highly theoretical exploration project. The assay results also surprised because they revealed shorter sections within the 38 metres averaging between two and three percent copper. Grades this high would normally be easy to see as shiny, metallic copper sulphides. How could it be that no copper was visible in RD1? In fact, the strange rocks from the bottom of RD1 had held an incredible secret since they were pulled from the Earth two months earlier. They were rich in chalcocite, a dark-grey to black copper mineral with a metallic lustre that is easily confused with hematite. Although chalcocite is a copper sulphide, it looks nothing like the brassy, brightly coloured sulphide of iron and copper known as chalcopyrite, which accounts for most of the world's copper deposits.

Haynes says they immediately went back outside to re-examine the drill core, this time more carefully with hand lenses. "We thought we could see some copper–iron sulphides, but we did not see the chalcocite, which contained most of the copper."

"We then returned to the office, where Dan got out his hand calculator, and estimated that the size of the 'hit' indicated by drill hole RD1 was of the order of 50 million tonnes, based on a block of mineralisation covering one square kilometre with drill hole RD1 in the middle of it! He then contacted Jim (Lalor) and Hugh (Rutter) with the great news."

The high excitement caused by the surprise results from RD1 is hard to imagine, but it was enough to get Lalor and Rutter on a plane that afternoon. Before the end of the day, they had joined Evans and Haynes in the backyard at Flagstaff Hill to see the core for themselves and begin new plans for exploration around Andamooka.

The assay results had profound implications. Haynes' model predicted that copper might or might not be in the Adelaidean sediments, but most certainly not in the rocks below. The host-rock part of the model had been thrown into doubt, although no-one was disappointed because it seemed they had almost snagged a copper orebody with their first drill hole. The results from RD1 also fuelled tremendous scientific curiosity; they seemed to have hit something beyond any of their wildest hypotheses. None of them had seen rocks like this at any university or in their study tours to virtually all of the world's major copper deposits. They certainly had no idea how this unexpected mineralisation had been formed.

The conflicting assessments of the basement rock highlight how far Western Mining's geoscientists were taken into the unknown by RD1. At first it was thought to be altered basalt, enriched with hematite, but this view had more to do with what they wanted to find rather than what really appeared before them. Basalt is a volcanic rock, erupted at the surface and quickly cooled from magma to a solid state. Minerals in volcanic rocks have little time to grow into crystals so basalt is fine grained. The unusual rocks from the bottom of RD1 could not be basalt because they contained some large crystals and abundant amounts of quartz. Lalor says David Duncan, a structural geologist who worked in Tim O'Driscoll's tectonics group, was the first to actually get it right. "We were all at Dan's office for the half-yearly reviews, and looking at

> **Chalcocite's rich reward**
> Chalcocite is normally a secondary copper mineral, produced from the alteration or break down of more common copper ore minerals. Haynes says there was some thought RD1's chalcocite was a secondary mineral, formed by the weathering of primary chalcopyrite ore on the ancient surface of the Gawler Craton "but we couldn't see the clay minerals that would be associated with significant weathering. It's actually a mineral known as digenite and is a primary copper mineral that occurs through large parts of the Olympic Dam orebody." Chalcocite has an atomic composition that makes it very high in copper metal by weight (79%), especially compared to Olympic Dam's other main copper ore minerals — bornite (63%) and chalcopyrite (34%). This was one another reason for the excitement caused by RD1's assay results. The high copper content of chalcocite would make it a focus of eventual plans for mining Olympic Dam.

the core from RD1. We were all thinking altered volcanics when Davy says: 'These are not altered volcanics — they're granites, they're altered granites!' We all poo-pooed him, but of course that's what it is. So, it was Davy Duncs who first called it."

It would be a long time before Duncan's opinion was accepted within Western Mining. Even in the 1980s, some of the company's top geologists were writing technical papers describing Olympic Dam as a deposit formed by sedimentary processes. For many years there remained a huge divergence of opinion on the genesis of Olympic Dam. There is still argument today, as described in chapter 3, but it is now commonly agreed Olympic Dam's mineral wealth is hosted by a vast subvolcanic breccia complex and likely formed just below a large crater lake in a particular style of volcano that occurs in rift valleys.

Beyond RD2

One of the biggest questions raised by RD1 was whether the very large and intense geophysical anomaly it drilled into was a ghostly map of a massive orebody. Answers were needed in a hurry, and they could only be found with more drilling. For each additional diamond drill hole, the company would need to find another $200,000 (in today's terms). This was a serious amount of money for Western Mining in 1975 following the collapse in prices for nickel, its main profit earner. Funds would need to be diverted from other activities, but there was no question of the need to sacrifice other exploration projects to fund follow-up drilling to RD1. The drilling strategy for the follow-up holes was hammered out in Lalor's office in Preston at a meeting attended by Lalor, Woodall, Evans and Rutter.

Two opposing strategies emerged from the meeting. One approach was to drill close to RD1 to maximise the chance of getting more copper and therefore some understanding of what controlled the location of the copper minerals — was it certain rock strata, was it fractures or faults? The second approach was to drill follow-up holes at large distances from RD1 on the assumption they were exploring very large anomalies and perhaps a huge body of mineralisation.

Evans says it was decided to drill RD3 and RD4 on a 400-metre grid, centred on RD1, as part of a "close in" drilling approach. The grid would progressively move north as a test of the alternative follow-up strategy of drilling at large distances.

On 27 October, 1975, Ted Whenan began drilling RD3 at a location 400 metres south east of RD1. The pace of drilling was accelerated by using a Schramm rotary air blast rig to "pre-collar" the hole to a depth of about 200 metres. The Schramm could quickly gouge a larger diameter hole through the top layers of Adelaidean sediments that were no longer of interest. Whenan's diamond drill would then come in and finish the hole, retrieving a core of the important rocks below at least 200 metres. The

Schramm rig was not only lower cost, it used compressed air and not precious water resources to lift the drilled rock fragments out of the hole.

A new source of water had also been found for the diamond drill holes; Tom Allison's generosity could not be expected to extend to unlimited access to his stock watering holes. The drilling crew on the Schramm rig had sunk several boreholes in the area, looking for a supply of groundwater that could be taken to the RD drill sites by road tanker. A fast-flowing supply of salty groundwater had been found about four kilometres west of RD1. The tanker would need to make two or three trips from the water bore to the diamond drill site during every 10-hour shift, but it was still far better than trucking water in from the nearest mains supply in Woomera.

> **Western Mining versus the opal miners**
>
> Lalor shared the news of the copper discovery with the director of the Department of Mines, Bruce Webb, on 2 October, 1975, less than two weeks after receiving the RD1 assay results. The briefing was unusually hasty, but Western Mining was eager to press its case for permission to drill in Andamooka's opal fields. The company's most favoured exploration target — ranking even higher than Olympic Dam — was an intense pair of magnetic and gravity anomalies at Bill's Lookout, inside the Andamooka Precious Stones Field. The Mining Act gave opal miners exclusive rights in this area. Before RD1, Lalor had raised with government the idea of exploring in the Andamooka opal field. Now that it had a stunning copper intersection at RD1, Western Mining had a strong argument for the creation of access, even if it meant an upset with the rowdy opal miners. On 8 October, 1975, Lalor wrote to Webb requesting access be investigated, and that the company's predicament be brought to the attention of the Minister of Mines, Hugh Hudson. Keith Johns, who was then deputy director of Mines, recalls today the Government considered amending the Mining Act to create a new form of strata title for exploration below 50 metres, but this was not pursued at that time. "The government met Western Mining's needs by cutting the size of the Andamooka Precious Stones Field in July, 1976. We took off the barren areas around the central diggings, which was enough to create access to Bill's Lookout. The opal miners protested for a while and formed the Andamooka Miners Association, but they were placated when we were able to do them a favour at nearby Stuart Creek. There was a 'rush' to these long-abandoned diggings, so we excised part of an Exploration Licence held by Amoco to create the Stuart Range Precious Stones Field." Amendments to the Mining Act were finally made in 1982 to allow access to Bill's Lookout, but the prospect was drilled without success.

RD3 made it through the entire layer of Adelaidean sediments, but the basement was unremarkable granite with no minerals of economic interest. Unlike RD1, the underlying rocks were not rich with dark-coloured hematite and not fragmented or brecciated. The team was getting colder. Hugh Rutter and Terry Brooks followed up RD3 with ground traverses, gathering more data on the gravity and magnetic anomalies in the hope of guiding the team to a more successful location for the next hole.

RD4 began a week before Christmas at a location 400 metres north west of RD1. The locations of RD3, RD1 and RD4 were on a straight line that traced the edge of the large gravity anomaly that drew them to Olympic Dam. The team still believed the odds of success were best on the edge of the anomaly, where localised changes in gravity were rapid. Before any drilling began, it was thought sharp changes in gravity could mark the edges of uplifted blocks of basalt — a crucial part of Haynes's model. RD1 had made redundant a large part of the original model, but a zone of abrupt change in gravity and/or magnetics was still a great target. These kinds of geophysical readings suggested strange things had happened to the rocks below, and just maybe this included one of the many processes that could concentrate disseminated copper into an ore body. RD4 made it through the Adelaidean sediments and found similar-looking rocks to those from the bottom of RD1, but without copper minerals. It was another barren hole. Lalor says there was no question of giving up because they still had no answers to the tantalising questions raised by the impressive copper intersection at RD1.

Western Mining had been in a similar situation a decade earlier, at Kambalda. The first drill hole intersected high-grade nickel sulphides. The next three holes were barren, but persistence carried them through to make a major discovery that would transform the fortunes of the company.

The "close in" drilling strategy had not discovered any copper mineralisation, and it had left the larger prospect area untested. Support grew for a switch to a larger grid pattern, but there was still resistance. No-one in the team had experience with exploration for large orebodies, even less so for those at great depth. The nickel orebodies at Kambalda were high-grade and confined to narrow shoots near the surface. The gold mineralisation of Kalgoorlie and surrounding districts was famously small veined. Experience in these environments was influencing the decisions to make small step-outs from RD1.

Rutter had modelled many hypothetical orebodies from the gravity and magnetic data. Many permutations of shape, size, density and orientation could account for the large and intense anomalies over Olympic Dam. While he did not know which model actually represented the rocks below the surface at Olympic Dam, Rutter was adamant the source of the anomalies must be very large. His most favoured explanation for the anomalies was an ancient volcanic centre — kilometres in diameter — because a pipe-shaped, heavy core of rocks was the best fit with the geophysical data.

It would prove to be an accurate interpretation, but the team was not ready to stretch its imagination quite that far. They settled on a compromise; RD5 would be drilled 1,000 metres north of RD1, some 2.5-times further out than either RD3 or RD4, and closer to the centre of the gravity anomaly. It was still not as far as the two kilometres that Rutter wanted to go northeast, but it was a step in the right direction.

The decision was rewarded when RD5, completed in the first week of March 1976, produced a 92-metre intersection of mineralisation at an average grade of 1.01% copper. It was still not an economic grade, but the thickness of the mineralisation gave the search an enormous boost. Woodall yet again rearranged exploration budgets across the company to fund another six diamond drill holes at Olympic Dam from within Exploration Division's budget.

Western Mining was also moving aggressively to lock up exploration title over many other geophysical anomalies in the district with similarities to Olympic Dam. On 31 January 1976, the company was granted Exploration Licences 231, 232 and 233, centred on the Bopeechee, Arcoona and Torrens prospects respectively. The three new licences increased the area covered by Western Mining tenements on the Stuart Shelf from 2,500 to almost 10,000 square kilometres. But the company was not just locking up areas to keep them out of the hands of competitors; even more exploration funds would soon be thrown into the Stuart Shelf to cover the cost of running a second diamond drill rig in the district.

As part of the build up of exploration activity, drilling supervisor John Emerson and his wife, June, moved home from Kalgoorlie to Woomera on 4 July, 1976. The date stuck in Emerson's mind because the young couple were greeted by the sight of Woomera's 1,500 US servicemen celebrating Independence Day.

The company judged it was still too early to make any public announcements. Western Mining's caution proved to be warranted. The next four exploration holes, from RD6 in March, 1976 to RD9 in September 1976, were bitter disappointments. RD6 was drilled 400 metres northeast of RD1, looking for a northern extension of RD1's mineralisation. When this hole proved to be barren, the team went back to their last success, RD5, and drilled RD7, 8 and 9 as 400-metre step outs to the east, north and west. The best result was only 14 metres of 1.2% copper at RD8. The grade was slightly improved from RD1 and RD5, but still not economic, and the 14-metre width of the intersection was a major step backwards.

DISCOVERY AT LAST

A year had passed since the surprise results of RD1 and interest in the project was beginning to wane. Evans had departed for Canada on Western Mining's study leave program to undertake a PhD at Queen's University in Canada. He was succeeded by Dave O'Connor, a Western Mining geologist with a great deal of experience in copper in Africa and Australia. Douglas Haynes, who had started it all, admits that even he was losing faith and had begun following up copper exploration targets in other states.

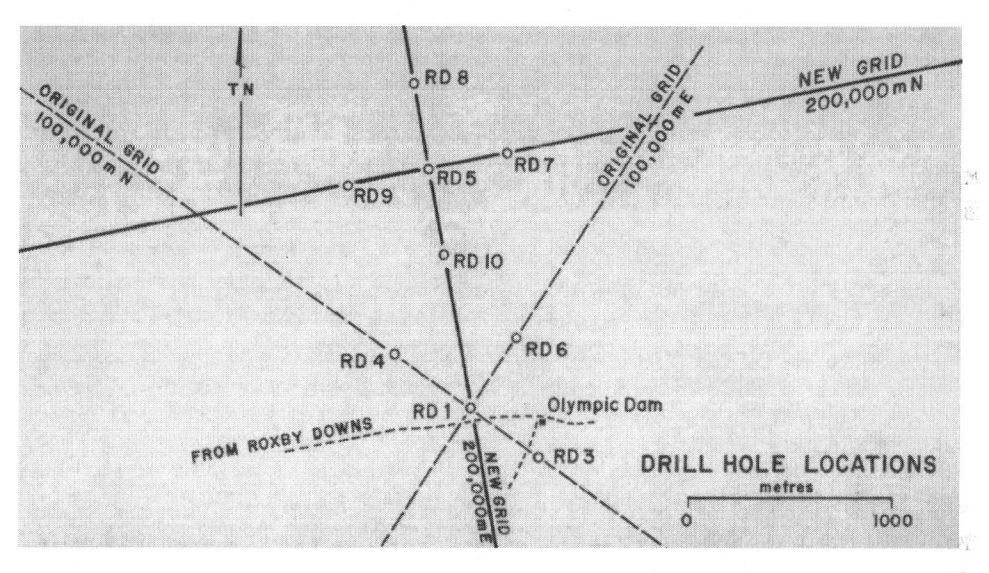

RD5 was a step-out on a new grid. It was rewarded with a promising intersection, but subsequent holes (RD6 to RD9) again disappointed. Source: BHP Billiton archive

The next few weeks would be curious, then extraordinary. Western Mining decided to make public the best results from its copper exploration on Roxby Downs station over the past year. The 1975-76 annual report, filed with the Stock Exchange on 26 October, 1976, stated that four vertical diamond drill holes near Roxby Downs station had intersected copper mineralisation at a depth of about 350 metres. The holes were spaced over a distance of about 1.5 kilometres and intersections varied from 8 to 92 metres at a grade of approximately 1% copper.

The statement was curious because it was still premature by Western Mining's conservative standards, especially when the most recent drill hole, RD9, was the fourth disappointment in a row. Pre-collaring of the next exploration hole, RD10, had begun on 21 September — about a month before this announcement — but there was no news yet on results from the deeper, diamond core drilling.

Western Mining's decision to make the announcement may have been influenced by rumours of a major copper discovery by a UK-based competitor, Selection Trust, at Myall Creek near Whyalla. It was said to be a stratiform copper deposit, the original target of Western Mining's activity. And rubbing salt into Western Mining's wounds, the discovery had been made on the southern Stuart Shelf, which the company now had a right to consider its own turf.

Selection Trust had been a large shareholder of Western Mining in the 1960s and, according to some accounts, had come close to launching a hostile takeover bid. The discovery of nickel at Kambalda in 1966 sent Western Mining shares soaring and was apparently all that saved it from falling into the hands of Englishmen. Given the rivalry between the companies, Western Mining would have been eager to stamp its authority by announcing copper discovery news of its own.

Then the extraordinary happened. On 5 November, less than two weeks before the company was due to hold its annual general meeting of shareholders in Melbourne, Arvi Parbo, Woodall, Lalor and other senior men from Western Mining touched down at the Roxby Downs airstrip to see some of the latest drill core. They had no reason to believe anything special would be produced by RD10, which had been sited midway between RD1 and RD5 — the only two holes that had produced anything worthwhile. In fact, the visit was a stopover on the way to their main objective, the company's oil exploration activity in the Pedirka Basin, many hundreds of kilometres further north.

An account of the sight awaiting them begins *The Olympic Dam Story* (see page 1). Drilling supervisor John Emerson was asked to bring two or three trays of core from the drill site to the side of the airstrip, about 30 kilometres away. When he arrived at the RD10 drill site with his Kalgoorlie-based boss Eric Steart on the day before the company's top men were due to fly in, Emerson found about 20 trays of drill core, twinkling with sulphides under the fierce brightness of the outback sun. Instead of the two or three trays that Lalor requested, Emerson took the lot. He was no geologist, but anyone who had been around diamond drill core could not mistake the primary sulphide ores. When Lalor, Woodall and others saw it the next day, they knew most of the core contained at least two percent copper, perhaps even three or four percent in some sections. It was the moment of discovery of Olympic Dam.

Lalor liked the experience so much that instead of getting on the two-way radio to tell the rest of the team, he engineered a similar surprise for Haynes. The young geologist was due to arrive in Adelaide in the next couple of days on a crisis mission. Selection Trust had announced its discovery and there was serious soul-searching within the copper exploration team.

DISCOVERY AT LAST

"I was thinking 'How the hell did we miss this thing?' I flew across to Adelaide to try to recover the situation. Obviously, the mood wasn't good. We had had been beaten to the punch by a competitor. In fact, I got a very distressed telephone call from Roy. He wasn't angry at me, but he was really quite angry that a competitor had found this copper deposit and, as he put it 'too many of our geologists are spending too much time in the office'. That's Roy, of course. He had that spirit; it was fair enough to feel that way.

"I flew in from Canberra for an early morning start and was quite down-in-the-dumps. I arrived and I was going to get a cab to the office in Marion Road. I had told Dave O'Connor not to worry about picking me up, but when I got to the airport, Dave and Jim were waiting. I said I wasn't expecting to see them, but Jim said he wanted to get my opinion on some drill core and that we would drive straight to the coreyard.

"When we got there, Jim said 'Go over and take a look at it to see what you think'. Of course, Jim and Dave knew all about it. When I saw the core it was the most astonishing surprise ever, actually — the highlight of an exploration career. A leader does things like that, and that was Jim."

Haynes says the next three days were amazing. "We had a lovely big dinner that night, drank too much red wine, but it didn't matter, it was an amazing time. We had found this enormous orebody. The moments after that were best described as ethereal. It was easily the best part of my career, and I thank Jim for that."

Haynes would have been even happier if he could have known that Selection Trust's announcement about Myall Creek would come back to haunt it. Further drilling showed the deposit was much smaller than Selection Trust had believed, and it remains undeveloped to this day.

Samples of the RD10 core were rushed to the Ballarat laboratory so that assay results might be available in time for the annual general meeting on 18 November. On the day of the meeting, Parbo, who was chairman as well as managing director, was able to tell shareholders that preliminary assays had confirmed the riches he had seen so unexpectedly next to the Roxby Downs airstrip less than a fortnight earlier.

Shareholders and the public were told that a new exploration hole, still in progress, had so far intersected 169 metres of mineralisation, starting at a depth of 348 metres.

"The upper 88 metres of the hole, for which accurate assays are available, averaged 2.41% Cu (copper), including 16 metres at 3.42% Cu, and 10 metres at 4.38% Cu. Provisional assays for the next 80 metres indicated an average grade of 1.9% Cu. The hole is still in mineralised material at a depth of 517 metres. It is clear that a very big body of copper mineralisation has been discovered, but a great deal more drilling will be necessary to establish the extent and the grade of the occurrence."

When completed, the assays revealed an average of 2.12% copper over an intersection of 170 metres. This was almost twice the length of the previous biggest intersection, but more crucially the average copper content was twice as good as anything found before. Rich grades above 2% could be the basis of a copper mine, even if it meant tunnelling half a kilometre underground to get to the ore.

The discovery of copper at Roxby Downs eclipsed all other news from Western Mining's annual general meeting that year. Parbo had plenty of other news for shareholders, including the likelihood of uranium production from its Yeelirrie project by the early 1980s and a more aggressive move into oil exploration. But financial and mining journalists knew an extraordinary drilling result when they saw it and led their reports of the meeting with the news of the copper discovery. It was the first time Olympic Dam made headlines. It was also the first time Western Mining talked about its innovative search for copper in South Australia's far north. In his chairman's address, Parbo said the exploration findings were the result of bold exploration in a previously unexplored and totally concealed area near Andamooka, following the development of new theoretical concepts. He also told reporters the company had spent $9 million (the equivalent of $45 million in today's terms) exploring for copper since 1958.

A uranium surprise

When all the assay results for RD10 came in they held another surprise for Western Mining; the hole had also found a very large body of uranium minerals of mineable grade. Assays of earlier holes had uncovered uranium at grades that made it a worthless and annoying impurity, but RD10 changed things dramatically. The same intersection that produced the spectacular copper result also assayed 170 metres at an average of 0.58 kilograms of uranium oxide per tonne. This was even higher than the average grade of the planned Yeelirrie uranium mine in Western Australia, although Yeelirrie would be a shallow, open pit mine with lower ore-extraction costs.

Western Mining had gone into the Stuart Shelf without any thoughts of finding uranium. It only began to assay for the radioactive metal after Hugh Rutter and Terry Brooks got a surprise result from lowering geophysical instruments down RD1.

This practice, known as a downhole geophyiscal log, allows mineral explorers to gather information about the rocks at short distances around the borehole. By extending the explorers' knowledge beyond the actual drill core, a geophysical log can provide additional clues in the search for economic minerals.

DISCOVERY AT LAST

The Herald, Thurs., Nov. 18, 1976

WMC FINDS COPPER IN SA

Big copper caps explor[ation]

WESTERN Mining Corp Ltd has discovered a high-grade, extensive copper deposit near Andamooka, in the Lake Eyre area of South Australia.

CHAIRMAN, Mr A. H. Parbo, told shareholders about the find at yesterday's annual meeting in Melbourne.

...the occurrence," he said.

The find is at the Olympic Dam prospect on Roxby Downs station, and caps an 18-year Australia-wide copper exploration effort by WMC costing $9 million.

Rule-of-thumb calculations put reserves at between 150 and [?]0 million tonnes [?]ly eight ver[?] drill

---

Western Mining Corporation Ltd. has discov[ered] grade copper over broad intersections near And[amooka] northern South Australia.

The chairman of WMC, Mr A. H. Parbo, [told share]holders of the discovery at the annual meetin[g ...] find caps a $9 million exploration effort.

Mr Parbo also told shareholders of the company's plans to:
• More than doub[le] treatment capacity of Kalgoorlie nickel smel[ter]
• Produce between [?] and 3000 tonnes of ura[n]ium oxide from the Y[eelirrie] little deposit in Wes[tern] Australia by the ea[rly] 1980s.
• Earn higher pro[fits] from its nickel min[ing] activ[ities]

Promising WMC copper find

THE FINANCIAL REVIEW 18-11-76

By JOHN BYRNE

WESTERN Mining Corporation Ltd yesterday announced details of what could be one of the most significant mineral discoveries ever uncovered in Australia.

Speaking at the annual meeting in Melbourne yesterday the WMC chairman, Mr A. H. Parbo, announced that a diamond drill hole in a remote area of South Australia had intersected 169 metres of copper mineralisation including 16 metres grading 3.42 per cent copper and 10 metres grading an extremely [?] per cent copper.

[?] mineralisation [?] 348 metres [?] at [?]crops.

MR PARBO

[an]nouncement was still 517 metres below surface.

The signif[ic]ance of the find is not only in the rich copper grades intersected in the hole currently being drilled, but also in the fact that the orebody has been uncovered in a totally concealed area with no surface [out]crops.

This indicates that WMC geologists have made a major breakthrough in exploration techniques.

Mr Parbo confirmed yesterday when he said t[he] discovery followed development of what [he] described as "new theore[tical] concepts."

Executives said yes[terday] that it was clear from [the] results that a very larg[e area] of copper mineralisati[on had] been discovered. "But," they ad[ded] a great deal more dri[lling will] be necessary to est[ablish the] extent and grade of the occurrence."

The discovery has been made on a pro[perty] as Olympic Dam

Modern geophysical instruments are highly sensitive and can sense far beyond the walls of the borehole into the surrounding rocks. They have been important in some recent discoveries, particularly in the nickel fields of Western Australia. But in the early 1970s, the downhole sensing technology was relatively new and geophysical logs were more of academic interest than anything else.

Rutter says the South Australian Department of Mines had made it clear it expected Western Mining to run geophysical logs of RD1 and RD2. After all, these were pioneering exploration holes that would be valued for their stratigraphic information, even if they did not find a speck of copper.

"We had our own geophysical loggers in Kalgoorlie, but they were all broken. So they sent across two loggers to Terry and told him to make something out of that. Terry was good, and he did. He took the bits from one and put it in the other until we had a gadget we could lower a probe down the hole and measure self-potential, point resistivity and natural gamma radiation."

Self-potential is the natural electrical current in rocks, while point resistivity is the response of the rocks to an induced electrical current. Both measurements say something about the electrical conductivity of the rocks. High readings can suggest the presence of copper sulphides, which are much more electrically conductive than regular quartz and clay minerals. Natural gamma radiation is the very high frequency electromagnetic waves from the decay of unstable atoms of uranium, thorium and even potassium. The downhole instrument is small, but is attached to a bulky reel of cable so it can be lowered many hundreds of metres to the bottom of the hole. With the logger and cable reel securely loaded on the back of their ute, Rutter and Brooks started the 1,300 kilometre drive from Melbourne to the RD1 drill location in South Australia's far north. Rutter says this trip was made not long after the first assay results had been returned to the Adelaide office.

"So we got across there and lowered the probe down. It was just a chart recorder with a wobbling pen. It got down to 335 metres and suddenly the needle darted across to the side and stayed there. We thought: 'Oh, it's broken!' So we turned it off, brought it all back up, looked at the probe. 'No, it's alright.' We put it back down again to the same place and off it goes the chart again. I said 'Let's change the scale' because we had it on the most sensitive scale. We put it on the most-coarse scale, and instead of going dead across to the other side it started moving backwards and forwards. I said: 'Terry, this is highly radioactive down there!'

"I didn't want to ring up Dan and tell him it was radioactive, perhaps uranium. I had already copped flak because the rocks we had encountered didn't look like basalts. So I went back to Adelaide and hired a spectrometer, a handheld gadget that measures potassium, uranium and thorium. I quietly took this up to Flagstaff Hill, ran it over the drillcore samples and lo and behold it wasn't potassium or thorium, it was uranium. It was anomalous in uranium, and seriously so!"

Rutter says he told Evans what he had found, and urged him to get the RD1 core assayed for uranium. Lalor recalls he was concerned by the results of the downhole logging and issued instructions to the team not to discuss with anyone the possibility of uranium at Olympic Dam. "From that early stage, I asked for all the drill holes to be assayed for uranium, with the strict instruction the results were to only to come to me at my home address (in Melbourne)."

Western Mining's Ballarat laboratory wasn't equipped for uranium assays, so its technicians were instructed to split the samples being tested for copper, gold and other metals and send a small part to AMDEL in Adelaide for additional, undisclosed tests. Up until RD10, the AMDEL tests had shown uranium to be no more than an annoying impurity. Lalor says he clearly recalls his reaction to the RD10 results. "I opened them up and looked at all the numbers. The deal was I was to call Arvi no matter what result came in. And I still recall it, I rang him and said: "Arvi, I think we have just ---ed up a good orebody!"

Lalor says that uranium would become an essential and highly profitable part of Olympic Dam, "but at the time it just wasn't something we wanted. I was particularly worried about the fact that copper and uranium were together. When that happens, you can't get all the uranium out of the copper concentrate. And it is a bugger to smelter. I knew right away we would not be able to smelt it anywhere. We would need to build a smelter of our own."

The uranium assay results from RD10 were announced on 11 February 1977 in the company's next quarterly update to the Stock Exchange. The uranium discovery attracted a small amount of news coverage; the media focus remained on the copper grades and how soon the company might know its ore reserves and mining plans for Roxby Downs. As usual for journalists, they were asking premature questions in the quest for their next story. In fact, the real headline waiting to be written in 1977 was that, while Roxby Downs promised to be a giant, it was still elusive.

10

A HUGE DEPOSIT REVEALED

The stunning result at RD10 led the exploration team to the reasonable view that drilling nearby should confirm a great discovery. In the parlance of exploration drilling, all they needed were some small step-outs. After all, they had found a 170-metre vertical interval of richly mineralised rocks; surely these rocks must extend in a horizontal direction over significant distances?

In line with this thinking, the exploration team planned to drill RD11, 12, 13, 14 and 15 in a tight group around RD10, with an average step-out of only 150 metres. But RD11 to 15 disappointed over the first six months of 1977. The first four holes after RD10 found copper and uranium mineralisation, but the best intersection of economic-grade copper was only one third the width encountered in RD10. RD15 yielded no significant results at all.

The new search for copper was now like a game of *Battleship*; the team knew there was an aircraft carrier out there, they just couldn't land a hit. In fact, drilling would later reveal that RD10 intersected a small outlier or pod of a very complex orebody. The margin of the main orebody was still two kilometres away in a north-west/south-east trending corridor. The pocket was so small that if RD10 were sited even 50 metres off its actual location, it could have been the tenth consecutive disappointment, not a stunning discovery.

At that point in the exploration program, the budget allowed for only one more exploration hole. If RD10 and then RD11 had failed, would Western Mining have given up the search? No-one could have criticised it; the company had already drilled perhaps twice as many holes as any other mineral explorer in the same circumstances.

Of course, RD10 found a mineral intersection that would keep them going. They didn't know how close they had come to losing Olympic Dam; all that mattered was the tiny window they had opened — a cylindrical drill core just 36.5 millimetres wide — had provided a glimpse of what looked like a giant orebody.

Project leader, Jim Lalor, says the results from RD10 were so spectacular there was never any question of giving up from that point, even after the disappointments of RD11, 12, 13, 14 and 15.

The inability to follow RD10 with another great intersection led to another rethink of drilling strategy. According to Lalor: "We had been messing around on the edges. We simply decided it was time to drill the whole anomaly, starting with a drill hole right in the centre of the target."

RD16 was drilled on the centre of the gravity high. It was a leap of faith that took them almost one kilometre northeast of RD10, but conservatism over the past six months had got them nowhere. RD16 found basement rocks even richer in hematite. They could have been mined as iron ore if they were close to the surface. It was confirmation that basement rock enrichment in hematite — a mineral that generates high gravity readings because of its above-average density — was the source of the gravity anomaly. But RD16 created a new set of problems; the drill rig struggled to power through the hard, dense rocks and the costly diamond-studded drill bits were wearing out faster than usual. Drilling supervisor, John Emerson, says that during the early holes at Olympic Dam, a diamond drill bit might normally get through between 15 and 30 metres of rock before wearing out. The hard, hematite-rich rocks required a new bit at least every four metres and sometimes a replacement was needed after drilling through just a single metre. Drilling progress was also slowed, with as little as two metres of core being pulled at the end of a 10-hour shift. The tough rocks at RD16 defeated them for the time being; drilling was stopped at 419 metres without intersecting any significant mineralisation.

Haynes says RD16 led him to postulate the idea of a large, barren core of hematite-rich rocks, with copper and uranium minerals in a zone around the periphery. If his "wrinkled annulus theory" was correct, drilling around the margins of the anomaly would provide the best chance of locating new zones of economic minerals. After Rutter checked his gravity data to confirm the locations of the anomaly's margins, the team agreed on a series of bold choices for the next drilling targets. The next hole, RD17 would take the search to the other side of the gravity anomaly for the first time and drill its outermost margin. RD17 would be 800 metres north of RD16 and 1.5 kilometres north northeast of RD10. This was no step-out drillhole — RD17 was more like beginning the search all over again.

The team's courage was rewarded with a 74-metre intersection of 2.49% copper and 0.45 kgs/tonne of uranium oxide. There was also more than 100 metres of weaker mineralisation below a depth of 401 metres. The copper grade from the latest hole was significantly better than RD10's average of 2.12%. In the world of mining economics, an improvement in copper grade of even 0.5 percentage points could mean the difference between an ordinary mine and a bonanza. And although RD17's mineralisation was less than half the thickness of that found at RD10, it was still a monster by any other standards.

RD17's results were announced in September 1977, but the news was overshadowed by other, quite dramatic events now engulfing Western Mining. The company was being hit by the full force of a global economic recession that began with oil price shock of late 1973. World demand for nickel — one of the major inputs in steelmaking — declined sharply and nickel prices fell by 20% between 1976 and the end of 1977. Western Mining was hard hit because it relied on nickel for its profitability. The gold mining operations recorded a loss in the 1975-76 financial year and income from the Darling Range alumina joint venture with Alcoa was still many years away.

The nickel market collapse was not unexpected by chairman Sir Lindesay Clark and other directors in the late 1960s. They had developed Kambalda at lightning speed because they knew the boom times in the nickel market would not last. Sir Lindesay also insisted on a conservative approach to financing a four-fold expansion of Kambalda and the construction of a nickel refinery because he believed some kind of bust would follow. Rather than borrow funds to develop the nickel operations, the Board asked shareholders to invest $45 million in new capital through a rights issue in 1968. At the time, it was a record capital raising by an Australian public company, equivalent to $450 million in today's terms.

Sir Arvi says the scale of that capital raising is hard to grasp today. "Correcting for inflation helps, but does not solve the whole problem. I well recall when an Australian company first made a profit of $100 million. This was at the time an unbelievable achievement. Today profits of billions and even tens of billions are commonplace. The difference is that, in addition to inflation, the scale of things has changed and, also, businesses are now much more capital intensive. Hand-held shovels have been replaced by mechanical loaders. You won't be able to find a shovel on a mine today; if you did, you won't be able to get anyone to use it! A similar change applies to everything."

But Sir Arvi says there is no difficulty understanding the significance of the rights issue for the company when nickel prices and demand in the 1970s fell further than anyone expected. "There is no doubt that decision by the Board — to raise capital rather than finance the initial nickel development with loans — enabled us to survive the subsequent world recession in the second half of the 1970s. Some of the directors wanted to borrow funds for the development, but Sir Lindesay convinced them it was too risky. He argued that Western Mining was exposed to high risks in its exploration spending and the demand and price for its products were beyond its control. It should not additionally expose itself to financial risk."

By mid-1977 the company had between four and five months of unsold nickel production sitting idle, and drastic steps were needed. In August 1977, Parbo announced that nickel production would be cut, and 600 workers would be retrenched at Kambalda — about 20% of the company's entire workforce.

These tough financial times continued for the next 18 months and would lead to budget and staff cuts right across the company. Lalor says he was faced with the difficult choice in 1978 of culling two or three senior positions in Exploration Division or laying off about a dozen geologists. He chose to make the senior positions redundant so he could keep the fabric of Exploration Division intact.

Exploration drilling budgets did not escape the cuts. The company might be exploration-driven and possess one of the world's most exciting prospects on Roxby Downs station, but going into the 1977-78 financial year there was just no money to maintain exploration drilling at past rates. The plan for the New Year allowed for only two or three holes, compared to seven or eight holes in the previous year.

To an outsider, it almost looked as though Olympic Dam was being put on the backburner, but the slowdown in drilling activity reflected an unhappy coincidence of events. Western Mining began exploring in the Andamooka region less than three years earlier when it was granted EL 190 in May 1975. The 2,358 square kilometre permit came with a commitment to spend $50,000 on exploration per annum, equivalent today to about $300,000.

By the end of 1977, the company had expanded its tenements to nine permits over a total area of more than 14,500 square kilometres, and was committed to annual spending of $3.5 million in today's terms. Nearly all of these commitments were made before anyone could have predicted the company's finances would be crushed by a global economic recession two or three years later.

The company's exploration commitments in the area were large, even for a major mining company in strong financial health. But two factors were behind its ambitious program to explore an area much larger than the ground around Olympic Dam. First, the region had many large and unusual geophysical anomalies like those at Olympic Dam. The science that led the team to Olympic Dam told them there could be other great copper discoveries waiting to be made in these locations. Secondly, their instincts as prospectors told them big discoveries usually came in groups of two or three; many of the biggest discoveries in the world were ore fields or camps, not isolated, single deposits. Everything they knew urged them not to leave the big geophysical targets elsewhere on the Stuart Shelf open for acquisition by competitors, even if it meant a near-impossible stretch of finances and manpower.

That's how 1978 began with a plan to drill only two new exploration holes at Olympic Dam, at a time when a new crescendo of activity could be expected. Funds for exploration drilling had to be shared across a number of active prospects in the wider region, including Bopeechee in EL 231, Andamooka Island in EL 233 and Arcoona in EL 232.

Fortunately for Western Mining, two more holes were all it would take to finally confirm the discovery of a major copper and uranium resource at Olympic Dam. RD18 was drilled in late 1977 and early 1978 at a location 800 metres east of RD16. It found only weak mineralisation, but holes RD19 and RD20 produced fantastic results and the first significant intersection of gold.

In May 1978, the company announced that RD19, 400 metres west of RD18, had intersected 13 metres assaying 2% copper, 0.03 kgs/tonne of uranium oxide and 15.4 grams of gold per tonne. This gold grade was four or five times the average of the company's underground mines around Kalgoorlie. Even if RD19 was a freakish gold intersection, it now looked like gold could be a sweetener to mining at Olympic Dam. It promised to be some compensation for the high costs of separating uranium from copper and the depth of the new resource.

Almost six months would pass before Western Mining announced two new drill results that finally gave it the confidence to state in the 1978 annual report that it had found a major deposit of copper and uranium at Roxby Downs, with the bonus of significant amounts of gold. The new drill results were RD20, 400 metres east of RD18, which encountered 130 metres averaging 2.8% copper and 1.1 kgs/tonne of uranium oxide between 450 metres and 580 metres. This had width comparable to RD10 and an even richer copper grade. The company also announced great results from deepening RD16 on the centre of the gravity anomaly with a more powerful drilling rig. The gravity high was not centred on a barren, hematite-rich core, but richly mineralised rocks. RD16 intersected 66 metres of 2% copper and 0.7 kgs/tonne uranium oxide between 666 metres and 732 metres. It was still in mineralisation when the results were announced on 18 October 1978. Haynes' theory of a barren core was not correct, but it prompted the team to take the big leap that was required to get from RD10 to the main area of mineralisation almost two kilometres away.

Western Mining also stated in the annual report that much more drilling would be required before the company could make any kind of estimate of the size of the deposit. It was as much as the company could reasonably say, but the results created almost unbearable anticipation in the financial and the wider community. It was known Western Mining had found very mineable grades of copper and uranium at drill holes in a scattered pattern at distances up to two kilometres apart; if economic grade mineralisation existed continuously over these sorts of distances, the Roxby Downs deposit would surely be one of the greatest mineral finds ever made in Australia.

Within weeks of the news, a prominent Adelaide mining engineer-turned-stockbroker, Norm Shierlaw, put his reputation on the line by taking out a full-page advertisement in the *The Advertiser*, Adelaide's broadsheet newspaper. He published his estimate that a mine at Roxby Downs could produce minerals worth $54 billion, or more than $200 billion in today's dollars. "If the whole project is developed to its logical conclusion, the benefit to Adelaide and all South Australians will be equal to if not greater than the potential advantage to Perth and Western Australians of the three billion dollar expenditure on the North West Shelf natural gas program. We know what the development of the iron ore province has already done for Western Australia and coal for Queensland."

History show's Shierlaw's estimate of the in-ground value of minerals was not bad, although it had more to do with luck than skill. It would take years of further drilling and many tens of millions of dollars before Western Mining would understand the size and nature of the deposit. Even three years later, in 1982, it was still coming to the realisation that while RD 17, 19 and 20 had indeed found the main orebody, the richest part was still waiting to be uncovered about two kilometres to the northwest.

There was still so much to learn about the deposit at Roxby Downs, but the project now belonged to others in Western Mining outside the Exploration Division. In 1979, Western Mining formed a new entity, Roxby Management Services, with the job of transforming the spectacular discovery into an orebody and a world-class mine. The Olympic Dam story would now be dominated by the task of finding a way to profitably develop a deposit that had no precedent anywhere in the world for its size, complexity and, of course, its uranium content. Western Mining had found a major copper deposit after 22 years of search. Now it must embark on a long and uncertain new journey to make the discovery pay.

⚒ 11 ⚒

Accidental elephant?

In his report on proposed drilling at Andamooka, Dan Evans wrote about the odds of "finding the elephant quite by accident". He actually used those words to argue the proposed program would not rely on luck or accident, but over the years some of the company's rivals have sought to dismiss Olympic Dam as a fluke and point out that even Western Mining admitted it did not find quite what it expected.

Analysing Western Mining's effort — in particular the conceptual thinking that led it to the middle of nowhere — is made more difficult by the fact there is still much controversy about the nature of the Olympic Dam deposit and how it was formed. But based on what we do know today, how does the Olympic Dam deposit stack up against the theories that took Western Mining to sink its first exploration hole on Roxby Downs station in 1975?

The team focused on the Stuart Shelf because it believed it would find hydrothermally altered basalts under a thick sequence of Adelaidean sediments. Furthermore, it was thought there was faulting in the region that could have uplifted blocks of basalt before sediments were draped over the top. This was an ideal setting for a large, stratiform copper deposit. Douglas Haynes' had proved with his PhD that altered basalts were potent sources of copper, and the model he developed suggested copper driven out of the basalt might be concentrated in certain sedimentary strata around the uplifted basalt blocks.

It was impossible to see uplifted blocks on the Stuart Shelf because of the thick cover of sediments, but basalt's distinctive physical properties could lead them to the right setting. An uplifted block of basalt would be associated with a gravity high because it would be composed of rocks of greater density than those around it. It might also coincide with a magnetic high because unaltered basalts at greater depths would produce a strong magnetic anomaly.

Western Mining also used the study of tectonics to help it locate a blind deposit of copper on the Stuart Shelf. It was believed that tectonic studies could locate major faults in the Earth's crust, and the intersection of these features would be a zone of structural weakness along which copper-rich fluids could migrate.

Drilling revealed the gravity anomalies were not produced by uplifted blocks of altered basalt, but by inverted-cone-shaped structures of altered volcanic rocks and altered granite. The alteration left these rocks rich in hematite, which is denser than the surrounding rocks and thus produces a sharply focused gravity anomaly.

The source of the magnetic anomalies is now thought to be massive hematite, a form of hematite that can be magnetic, as well as magnetite-rich rocks at depth — below the zone of hydrothermal alteration.

Some might argue it was fortunate the Olympic Dam deposit produced geophysical anomalies that looked like they could have been generated by uplifted blocks of basalt. But the written records of the exploration team make it clear that Olympic Dam would have been drilled even if the anomalies did not fit the model. Hugh Rutter wrote to Douglas Haynes and others saying the geophysical anomalies were not a great fit with the model, in fact they seemed to him to be generated by ancient volcanic centres. This was precisely correct, which has only been understood in more recent years. Despite the mismatch between the model and the geophysics, Rutter and others wanted to drill because of their scientific curiosity. The anomalies were large, intense and mysterious; something strange must have happened to the target basalts that were thought to lie beneath the deep cover of sediments.

Of course, sheets or piles of basalt were not found, but there was enough copper to make Olympic Dam the fourth largest copper deposit in the world. So Haynes was correct in predicting copper would be in the area.

It is now accepted basalts that eroded off the surface of the Gawler Craton before deposition of the Adelaidean sediments were indeed a substantial source of copper. Some remnant basalts were found during the initial drilling out of the resource, and recent drilling by BHP Billiton has found more significant thicknesses in the order of several tens of metres.

Haynes' model also correctly predicted that hydrothermal fluids could transport the copper into other environments. He was wrong about where to find the trap for that mobilised copper or, in other words, the type of rock that would host the copper deposit at Olympic Dam. But two out of three correct predictions got the company across the line, with persistence and a dash of good luck.

Haynes acknowledges the host rocks and the style of mineralisation were different to what the team expected. "Clearly, it was. The style of mineralisation we were seeking at Olympic Dam was similar to Mt Isa copper. But the source rocks were the same — these altered basalts. It's just the manner of precipitating copper is different. In the case of Mt Isa, the copper is from the reduction of an oxidised hydrothermal fluid. At Olympic Dam, it is oxidised fluid reduction by fluid mixing in a volcanic setting. But the source rocks are the same and it's the source rocks that drove us into the Olympic Dam area. That's the crucial point about the discovery. If we had focused on host rocks, such as those hosting the types of deposit Mt Isa, I don't think we would have found Olympic Dam, not as early as we did anyway."

The discovery story has other twists. It's unclear now whether the horizon identified by the Department of Mines experimental seismic survey was indeed the boundary between the basement rocks and the overlying sediments. At the time, the news of a major horizon at a relatively shallow depth of 330 metres led to the belief that drilling at Olympic Dam should reach the basement and resolve the question about the presence of altered basalts. Subsequent drilling and geophysical surveys showed this horizon actually may have been a contact between two sedimentary rock units within the lower sections of the Adelaidean sediments. If so, it was just a stroke of luck that it looked like a basement horizon and provided the team with greater confidence about the location. There's no clear answer about this because the results of the experimental seismic survey were never that clear. But it's possible the survey was much like Rutter's find of a malachite speck on his first ground traverse; in the end it wasn't a material fact, but it greatly added to the team's belief in what they were doing and boosted their confidence about drilling at Olympic Dam.

The contribution to the discovery of tectonic lineament studies by Tim O'Driscoll and David Duncan continues to be a source controversy. Roy Woodall says these studies were critical and their importance is overshadowed by the geophysical aspects of the discovery history.

The leader of the Olympic Dam exploration project, Jim Lalor, says lineament studies played a part in identifying Olympic Dam as an exploration target. "There's a wide range of views about the importance of tectonics in the discovery. It was a factor, and at the time it was certainly one of a number things we drew upon that gave us confidence about where we were looking."

Lineaments played a role in the discovery, regardless of whatever views the geoscientific community held at the time about this type of study or indeed those views today.

Haynes says opinion against tectonic studies has hardened in recent decades for two main reasons; outcropping rocks above major lineaments don't show the expected geological structures, and lineament theories are largely inconsistent with everything we do now accept about plate tectonics.

"While some lineaments of the style defined by Western Mining's tectonics group might truly reflect deep structures in the Earth's crust, there appears to be no way of distinguishing these from random linear patterns."

In other words, some lineaments are merely random artifacts of nature, just like the face of a famous person appearing on a piece of toast. They cannot be differentiated from lineaments that might actually mean something.

Haynes says studies of regional geological structures around Olympic Dam have not helped the cause of lineament studies. He points to recent geophysical imaging in three dimensions by Geoscience Australia and PIRSA[13] that did not find evidence of the two major lineaments identified at Olympic Dam by Western Mining's tectonics group. The study, based on regional gravity and magnetic data and a 200-kilometre seismic survey that penetrated to depths of more than 40 kilometres, found sets of regional structures trending northwest and northeast, but not the north-northwest structures that dominated O'Driscoll's studies.

In addition, underground mining of the deposit over many square kilometres in the past 22 years has not encountered the geological structures defined by Western Mining's tectonics group, although the orebody does extend along a north-northwest orientation that appears to fit with O'Driscoll's "tectonic corridor".

Recent drilling and research promises to shed more light on the subject. Kathy Ehrig, principal geometallurgist of BHP Billiton Uranium, says recent drilling has provided detailed information for the first time across the entire orebody and the areas around it. She says structures are now clearly visible either side of the orebody, which are not seen in the central part of the Olympic Dam deposit because of ore-forming processes. Ehrig says a major structure runs through the orebody, along which there is elongation of the mineral resource, and recent work has also clearly shown a chemical signature it has left behind.

---

[13] *Regional crustal setting of iron oxide copper-gold systems of the Olympic Dam region: Insights from potential field modelling*, a paper published in the Journal Economic Geology (2007) by N. Direen and P. Lyons from Geoscience Australia. (2) *The 2003 Gawler Craton Seismic Survey*, a report by P. Lyons and B. Goleby published by Geoscience Australia (2005). *(3) Geodynamic setting and controls on iron oxide Cu-Au (U) ore in the Gawler Craton, South Australia*, a paper by N. Hayward and R. Skirrow published in a volume edited by M. Porter (2010).

Persistence

An analysis of the factors in the discovery would not be complete without looking at the role played by persistence. All exploration projects need persistent effort, but it was an especially large factor on the Stuart Shelf where there is no rock outcrop to explore and understand. In many ways, the Roxby Downs exploration program was more like a search for oil and gas than a conventional hunt for minerals; it relied on geophysical information to pinpoint the prize in unknown rocks hundreds of metres below the surface. Oil and gas explorers know the odds of success in these conditions are small; mineral explorers expect results with fewer drill holes.

There were a number of occasions, particularly before RD10, when Western Mining could have walked away without disgrace. It is difficult today to appreciate the importance of persistence in the discovery because we now know a great discovery awaited the team. At the time, these explorers had no idea about the future. They had to fight against limited budgets, the demands of competing exploration projects, of which there were many, and their own creeping doubts when drill hole after drill hole came back with disappointing results.

It's impossible to know what might have been if Western Mining's exploration team was not so determined and was not part of a company that gave its geoscientists so much freedom. However, it is instructive to look at the exploration success record of other companies on the Stuart Shelf since Olympic Dam. Explorers from around the world flocked to the region after the announcement of the results at RD10. Most departed within a few years without success. Only two major copper discoveries have been made in more than three decades since Olympic Dam — Prominent Hill and Carrapateena.

Prominent Hill is located 180 kilometres northwest of Olympic Dam. It is a major discovery, although not the same order of magnitude as Olympic Dam. The deposit was narrowly missed in 1990 by Metals Exploration and Burmine, a company headed by South Australian geologist, Derek Carter. A drill hole just 100 metres from the ore body returned a low grade of 0.2% copper over intervals of several metres in magnetite-rich rocks. Burmine ran out of cash and opportunities to complete its planned drilling program, but Carter persisted. Over the next 10 years he formed a new company Minotaur Gold (later renamed Minotaur Resources) and doggedly pursued a new partner, Billiton (prior to its merger with BHP) with the cash and the desire to support another drilling program. A critical decision was made to focus drilling on the local gravity anomaly, not the magnetic anomaly as previously targeted. The deposit was finally discovered in late 2001 by URAN1, the first drill hole on a prospect known as Uranus. The name was promptly changed to Prominent Hill when Carter realised he would be remembered as the man who discovered Uranus.

As an interesting footnote, Billiton merged with BHP in 2001 and sold its 50% interest in Prominent Hill because the project was deemed to be too small. The current owner of the mine, OZ Minerals, began earning an interest in Prominent Hill from Minotaur in 2003 and eventually took over the entire project in an agreed deal. Minotaur's shareholders ended up with shares in OZ Minerals, which has allowed them to share in production and benefit from further discoveries, which look increasingly likely.

Carrapateena is a more recent discovery, uncovered in 2005 by RMG Services on the western edge of Lake Torrens and only 70 kilometres southeast of Olympic Dam. The Adelaide-based private company takes its name from the initials of the company's founder, Rudy Gomez, a Philippine-born engineer, metallurgist, inventor and part-time prospector who is aged in his 70s.

The size of the Carrapateena discovery has yet to be announced, but there is speculation it could rival Olympic Dam. Canadian company Teck Cominco has conducted a massive drilling program as part of a deal to earn an interest in the deposit. The most exciting drill intersection to be announced was in 2007, when Teck revealed it had intersected 905-metres averaging 2.08% copper and 0.92 grams/tonne of gold.

The Carrapateena story is also one of persistence. Gomez took 16 years to drill the prospect, and only then by gambling his retirement savings on the cost of sending a drill rig to the remote location. South Australia's Department of Primary Industries and Resources provided a grant of $99,175 through its PACE program to cover part of the drilling costs, but with a condition the funds must be spent within 12 weeks. The timeframe meant the only available driller was a company that usually sunk water bores in Adelaide. But the rig made it all the way to Carrapateena and after three attempts reached its target depth of 500 metres, where the discovery was made.

Gomez says a major reason for his persistence was advice from Tim O'Driscoll, the leader of Western Mining's tectonic studies group. "I first met Tim in Adelaide in 1979 where he gave me advice on structures regarding my oil exploration in the Philippines, and again in 1996 after Tim retired from Western Mining and was consulting for the oil division of the SA Department of Mines. He told me the Carrapateena arm of Lake Torrens was a very important structural feature. That advice was one of the main reasons I refused to give up on drill testing the anomalies we generated over the eventual discovery."

The stories of both Prominent Hill and Carrapateena reinforce the importance of persistence in locating blind or deeply buried copper deposits on the Stuart Shelf. The fact that only two discoveries of copper have been made in more than three decades since Olympic Dam has also enhanced Western Mining's reputation as a bold and innovative explorer.

12

THE QUIETLY ACHIEVING PARTNER

Norm Shierlaw's full-page advertisement in *The Advertiser* in January 1979 created extraordinary excitement about the economic boost South Australia could expect from the mineral discovery at Roxby Downs. It was easy to get carried away with his conclusion of a resource containing some $200 billion in mineral wealth, especially in a state in dire need of major projects to boost economic activity and cut high levels of unemployment.

Behind the scenes, Western Mining's leaders were having far more sober thoughts about the reality of bringing Olympic Dam into production. The first challenge was to "drill out" or define the deposit by sinking diamond drill holes on a closely spaced grid as tight as 50 metres by 50 metres.

The known area of mineralisation was already in the order of 10 square kilometres; "drilling out" a resource to determine its position, grade and continuity with enough confidence to invest in a mine would require several hundred holes over a number of years. A drilling program of this size would cost well over $100 million in today's terms — a daunting prospect for a company that a year earlier could budget for only half a dozen holes over all its tenements on the Stuart Shelf, including just two drill holes at Olympic Dam.

These costs were just the first stage of developing Olympic Dam, which would ultimately require the investment of many billions of dollars, much of it years in advance of the sale of a single pound of copper. The company's exploration success far exceeded its balance sheet or its financial capacity to develop what it had found. Western Mining had searched for decades for a big deposit of copper, but it had not counted on a deposit complicated by uranium or buried as deep as Olympic Dam.

The cost of separating uranium and removing the last trace of radioactivity from the copper, gold and silver added to the financial challenge. The combination of such concentrations of uranium with copper and gold had not been encountered anywhere else in the world.

Western Mining knew at an early stage it would need a partner. In December 1977, almost a year before a major discovery was confirmed, Arvi Parbo told shareholders in his annual address "many companies have indicated their interest in joining Western Mining Corporation in this project. Because of its likely size it is realistic for us to think in terms of eventually having partners, and in due course we will be having discussions with those who have indicated interest."

The rush to join Western Mining at Olympic Dam was welcome, although the company's leaders had mixed views about joint ventures. According to Hugh Morgan, who led Western Mining as managing director from 1986 to 2003, the company had experienced problems in its joint venture in the Koolanooka iron ore mine in the early 1960s. This had been a factor in its determination several years later to develop Kambalda in its own right.

But the company was also pragmatic. Sir Arvi says today there was no realistic possibility for Western Mining to develop Olympic Dam on its own. "Once drilling had indicated that this was ultimately going to be a large project, there was no question in my mind that we had to have a financially strong company as a partner. We knew that they would be much more bureaucratic, would look at many aspects of the project differently from us, and that there would be problems, but this was something that had to be accepted. We had to make sure that WMC was not put into financial jeopardy if things went wrong."

Sir Arvi says that by December 1977, the Board of Western Mining had already discussed and sanctioned the development of Olympic Dam as a joint venture. "Sir Lindesay Clark was still a director and I do not recall he having any different views. Had he questioned it I would remember this, but in my recollection the Board was unanimous."

Sir Arvi also provides a glimpse inside the Western Mining boardroom, which was as unconventional as many other aspects of the company. "We never voted at Board meetings. Decisions were made by consensus. If it was a complex issue I would, after a full discussion, outline what I thought the consensus was and ask if any director disagreed. If there were serious concerns, we would defer the decision and discuss it again, both informally between meetings with the directors concerned and at the next Board meeting, modifying the decision if necessary until their concerns were met. In my time this always worked."

BP was an early and eager suitor. Sir Arvi says the Australian chiefs of BP's minerals business had made an approach some time in 1977, saying BP was keen to join Western

Mining on a number of its projects, not just Olympic Dam. It stepped up the courtship with an invitation to Parbo to stop over at BP's London headquarters on his way to Estonia for Christmas at the end of 1977.

The first contact in London was with Frank Rickwood, an Australian who had risen to the top of BP's global coal and minerals business. "Rickwood explained the view, then universally held by oil companies, that oil will gradually phase out and BP must acquire other interests to take its place. They wanted to explore opportunities for cooperation with WMC ranging from joint exploration (such as in the Benambra Project) to farming into discoveries such as Yeelirrie and at Roxby Downs, to becoming a shareholder in WMC. Our Exploration Division had a very high reputation and record of success and this was an area where they needed assistance.

"I explained that we were unable to talk about Yeelirrie because of exclusive discussions with another party, unless these discussions were not successful. WMC would be interested to consider, without commitment, any other opportunities for cooperation, although the possibility of BP becoming a substantial shareholder in WMC was unlikely to be acceptable because we would lose the advantage of our Australian character. Rickwood repeated their strong interest in discussing a possible farm-in to Olympic Dam."

BP's interest in Olympic Dam went right to the top of the company, with BP's deputy chairman, Monty Pennell, and managing director, Robin Adam, telling Parbo at a lunch on the same visit to London of BP's "very strong" interest in the new copper and uranium discovery in the far north of South Australia.

At this stage, Western Mining's most recent exploration success at Olympic Dam was RD17, for which results were announced three months earlier in September 1977. The announcement of subsequent successes — RD20 and a deepened RD16 — and public confirmation of a great discovery would not take place until October 1978.

But as early as April 1978, Rickwood visited Australia in a bid to begin negotiations. BHP was also keen to get involved, with BHP's chief general manager, Brian Loton, calling in to see Sir Arvi in August 1978 about a possible joint venture.

In mid-1978, Western Mining had formed a team under Hugh Morgan to conduct a global tender to secure a joint venture partner on the best possible terms. The team included Jim Lalor, finance director Don Morley and in-house lawyers Kym Saville and Colin Wise. For Morgan, it was the beginning of 25 years of involvement in Olympic Dam, which he describes as the highlight of his career. Morgan would go on to become managing director as well as a leading figure in Australian business, a Reserve Bank director and President of the Business Council of Australia, but none of these achievements topped the experience of bringing Olympic Dam into production.

The team's first objective was to define the minimum requirements of a joint venture partner. Money was at the top of the wish list, in particular the capacity to loan Western Mining the funds it needed for its share of the high costs to develop Olympic Dam. By the time Sir Arvi met BHP's Brian Loton, he was able to outline the following terms for a joint venture:

1. A cash premium of $5 million ($21 million in today's terms)
2. Partner to carry costs of drilling and metallurgical work in Stage 1 estimated to total $30 million ($128 million), which would earn the partner an 18% interest in the project
3. Partner to carry further costs of $20 million ($85 million) in Stage 2 to complete a feasibility study, which would increase the partner's interest to 49%; and
4. The partner to propose how it would assist Western Mining finance a 51% share of the development and construction costs, which would allow the partner to increase its equity again up to 49%.

"Although it was not possible to be specific at this stage, I said I thought the project costs would be of the order of $1 billion ($4.3 billion)."

The joint venture team inside Western Mining also drew up a list of possible partners, which included most of the international giants of the oil and minerals industries.

Morgan says Western Mining's search for a partner gained urgency with the announcement of the discovery in October 1978. "The retired head of South Australia's Department of Mines, Sir Ben Dickinson, wrote to the Premier, Don Dunstan, urging him to nationalise the Olympic Dam discovery. He wrote that Western Mining would not have the capacity to bring it into production, so the Government should take control of this important state asset."

Morgan says Sir Ben was highly influential in South Australian politics and the possibility of an Australian government taking control of Olympic Dam was not as outlandish as it sounds today. As recently as 1974, the Whitlam Government had declared it would take control of the marketing and export of all Australian uranium, and would also take charge of all exploration for uranium in the Northern Territory.

"Sir Ben's letter to the Premier certainly put a fire under our efforts to find a joint venture partner."

Norm Shierlaw was also pressing for at least some State Government ownership of Olympic Dam. He stated the Government had established a precedent by taking an effective 18% interest in the Cooper Basin through the South Australian Oil & Gas Corporation. He calculated that a comparable equity interest in Olympic Dam, plus the royalties from the entire project, would cover about one quarter of the state budget.

Sir Arvi also recalls Sir Ben urging the South Australian Government to intervene in the development, although he was not so concerned the advice would be followed. "Ben was muddying the waters ... but in a sense it didn't matter. Don Dunstan was dead against uranium. They were at loggerheads. Dickinson wanted uranium developed, but by the state. Dunstan didn't want it developed at all."

Within days of the October 1978 announcement, Morgan and Morley began a global tour to sound out interest from BP, Shell, Mobil, Texaco, Exxon and Anaconda. Morgan says they went with a list of things Western Mining wanted from a joint venture partner and asked interested companies to put forward their best offer by 13 December 1978. Eight companies submitted bids, including BHP and Utah, but at the end of the tender process, BP was still the most eager potential partner. Morgan says BP was the only bidder that agreed to all of Western Mining's major terms, including acting as banker for Western Mining's 51% share of the development costs. Furthermore, BP agreed to secure funding against the project — without recourse to other assets of Western Mining — and repaid only when the project began to generate cashflows.

The funding arrangements were a crucial part of the deal because banks were reluctant lenders to the boom-and-bust resource industry at the best of times. When uranium was involved, the odds of a bank loan were even longer because of the conservatism of the electricity utilities that bought uranium, as Morgan explains. "An iron ore buyer will give you a contract. You can then go to your bank and say 'I have sold it' and arrange finance to develop your iron project against that. Uranium buyers will not enter a contract until the mine is built and you can actually deliver your product. There was this long flirtation that went with visiting and setting up friendships with utilities, where you got a wink and a nod that 'yes, we will deal with you, but we can't actually formally deal with you'. That was the case in Korea, Japan, the United Kingdom and France."

The terms of the joint venture were finalised by mid-1979, with Morgan leading the negotiations for Western Mining and Rickwood for BP. The deal was announced in July 1979 to the applause of media commentators, who judged it an outstanding deal for Western Mining. Olympic Dam was always going to attract an impressive field of suitors because of its world-class status, but Morgan and his team had certainly wrung every advantage from the company's great mineral discovery.

In summary, BP agreed to pay $5 million ($21 million in today's dollars) at the start of the joint venture, to provide $50 million ($210 million) to meet the estimated cost of exploration, metallurgical testing and other work to complete a feasibility study, to finance Western Mining's share of the costs of developing the mine and processing plant, and to spend $10 million ($42 million) on exploration in Western Mining's vast tenements across the Stuart Shelf over the next three years.

The decision by one of the world's largest companies to become a partner in Olympic Dam had some other positive consequences for Western Mining. It silenced Sir Ben Dickinson and anyone else who doubted Western Mining could muster the capacity to develop such a large and complex resource.

BP's presence also increased the pressure on the South Australian Government to drop its ban on uranium mining. With development now a genuine prospect, the Government would be forced to choose between its anti-uranium principles and prosperity.

Morgan says the joint venture began with a very positive spirit, but this soon changed. "The problem was, after the deal was agreed, responsibility passed from Rickwood and his minerals people to the finance department. The minerals group was a recent addition to BP through the acquisition of Selection Trust and some other companies. They really had a different culture to BP. They were miners like us and we got on well with most of them, despite the fact that Selection Trust had been a fierce rival of Western Mining before the BP takeover.

"However, the finance department was a different story. It was steeped in BP culture, highly bureaucratic and it wanted a much greater involvement in managing Olympic Dam, even though Western Mining was clearly the operator under the joint venture agreement."

Morgan says one of the biggest sources of tension was the fact Western Mining agreed to assign BP a 49% interest from the beginning, rather than progressive step-ups in its percentage interest as money was spent.

"When we negotiated the joint venture, BP pointed out that each time it increased its equity, it would need to go back to the Foreign Investment Review Board for approval. It didn't like this arrangement because of the risk it might be refused at a later stage. We accepted the argument and granted 49% up front, with the money to follow. But over time, people forgot all of this. What began as a logical request created a problem in the finance department. It thought 'We have a 49% interest so why are we putting up 100% of the money?' There were a lot of hesitations and tensions."

In many ways, BP and Western Mining could not have been further apart in their approach to doing business. BP was heavy on process and committees, while Western Mining saw its lack of formality and quick decision making as some of its main competitive advantages. The companies settled into an uneasy partnership for the sake of getting Olympic Dam developed, even if would take much longer than Western Mining had anticipated.

BP's approach to the joint venture is well described in an unpublished account written for Sir Arvi by Kym Saville. He records there was a serious debate between Western Mining and BP in 1984 on the terms of reference for the feasibility study to determine whether Olympic Dam should be developed.

"Under the terms of the Joint Venture agreement, the terms of reference for the feasibility study had to be agreed by the parties (with disputes referred to an independent expert). The Manager (Roxby Management Services) was then to undertake the feasibility study in accordance with the agreed terms of reference. If in the opinion of the Manager (which opinion could also be referred to an independent expert, if BP disagreed) the feasibility study demonstrated the project was technically and commercially viable, then BP had six months following completion of the feasibility study to commit to the project, failing which it was deemed to have withdrawn.

"BP was conscious of these requirements and anxious to avoid being placed in a situation where it either had to commit (and finance 100% of) a project it did not like, or lose its entire interest. Conversely, WMC was determined not to allow BP to unduly delay progress of the project. When, in early 1984, WMC proposed terms of reference for a feasibility study, BP sought to include a large number of financial and marketing tests of commercial viability. These included not just minimum rates of return, but a range of sensitivity tests, debt servicing capacity tests and a requirement that long term sales contracts be concluded for a minimum percentage of copper and uranium production. Even the most robust project would have had difficulty meeting all of these tests, and the small scale initial project at Olympic Dam was never considered likely to be robust. Rather, it was seen as a logical first step on a long road to the development of a much higher optimum capacity. WMC saw in the proposals the prospect of long delays with commencement ultimately being deferred until BP was ready to proceed. If BP could attain such a position, there was also a concern that it would use its power over the timetable as a bargaining chip to negotiate a better deal. (BP had already succeeded in revamping some of the financing provisions in 1982, as a result of which the interest rate paid by WMC on BP project loans increased by almost 2% per annum).

"A long debate ensued, culminating (as it often did with BP) in crisis negotiations in Melbourne with Ted Hannington and Hugh Morgan which went all night so they could be completed before Ted returned to London the following morning. I recall a surreal experience after we had given BP a final proposal about 4 a.m. and were waiting for them to get back to us. Hugh, Jeff Smith and I were sitting in the meeting room on the 29th floor at 360 Collins Street, watching a terrible Swedish movie on SBS, all of us half asleep, waiting for the phone to ring. When finally it was agreed, I had to drive to Grahme Dixon's (the company secretary's) house and wake him up at 5.30 a.m. to sign the agreement before driving to the next suburb to similarly rouse the BP company secretary so that Ted could take the signed agreement with him to the airport!

"In the end, we were able to compromise in a manner that removed all of the prescriptive tests and ensured that BP would not unduly delay the process. Nevertheless, as you will well recall, when the commitment decision was finally upon them in late 1985 they tried very hard to defer it."

On 8 December 1985, the joint venturers advised the South Australian Government they would develop Olympic Dam to produce 55,000 tonnes of copper, approximately 2,000 tonnes of uranium oxide and about 90,000 ounces of gold per annum. Construction would begin in March 1986, with first production expected by mid-1988.

The difficult, 14-year relationship with BP came to an end in 1993 — five years after Olympic Dam was commissioned — with Western Mining buying BP's share of the project. The cost of the buy-out in 1993 dollars was $US240 million plus a payment of $US190 million to repay the balance of BP's loans to Western Mining. In current dollar terms, the total cost for BP's stake was about $US650 million. Today this seems like a remarkable bargain for half of the world's greatest mineral deposit, but the price was struck when Olympic Dam was only a fraction of its current size and in an era of depressed commodity prices. The China-led resources boom was still more than a decade away.

The sale price was determined by a deal BP had recently reached to sell its share of Olympic Dam to Minorco, a South African-based miner. As with most joint venture agreements, Western Mining had what is known as a pre-emptive right or first right of refusal over BP's stake if BP wanted to sell. This means Western Mining could choose to bump Minorco as the buyer as long as it matched the terms of the sale, which it did.

It was BP's second attempt to exit the Olympic Dam project following a change of strategy in the late 1980s to refocus on its energy business. The first attempt in 1989 was dropped after a legal challenge by Western Mining. BP had reached an agreement to sell its 49% interest in Olympic Dam to Rio Tinto as part of a multi-billion-dollar, global package of mineral assets. The sale price for BP's Olympic Dam stake was way above any rational assessment of its value, which put Western Mining in the invidious position of either paying perhaps double the fair market price to become the sole owner of Olympic Dam or swapping BP for Rio Tinto, another London-based resources behemoth. Kym Saville says Western Mining issued proceedings in the Supreme Court of Victoria, challenging the validity of the pre-emptive rights notice. The dispute continued until a few days before the scheduled trial date, when BP withdrew the notice and removed Olympic Dam from the sale agreement with Rio Tinto. "Although we had doubts about the veracity of the purported sale price, the primary legal ground of our objection was that BP was seeking to transfer its joint venture ownership interest to Rio Tinto, but to retain the existing project loans and the residual obligation to finance Western Mining's share of the cost of future expansions up to 150,000 tonnes per annum of copper. It was our view that the financing obligations were inseparable from the joint venture interest, and we were anxious not to put ourselves in a position where BP Finance acted purely as a financier without any shared interest as a co-owner of the project. BP was forced to frame its offer on this basis as Rio Tinto refused to take on the financing obligation."

It was the lowest point of the relationship between Western Mining and BP and undoubtedly strengthened Western Mining's resolve to see off its difficult partner when the chance arose with Minorco a few years later.

It is now almost 20 years since BP quit Olympic Dam and the minerals business worldwide. At the time, it had an advertising slogan about being "the quiet achiever". BP executives in London who visited the Olympic Dam site in the late 1980s were shocked to see Western Mining engineers and geologists wearing T-shirts that parodied the slogan with the new tagline about BP quietly achieved not very much all. Western Mining workers were venting their frustration with the endless delays and bureaucracy of BP, but in the end the British oil giant lived up to its advertised promise. It provided the financial backing to develop Olympic Dam when other partners and Australia's banks were not prepared to do so. It made a critical contribution that is largely unrecognised today.

QUIETLY ACHIEVING PARTNER

13

DRILLING OUT

With the arrival of BP in July 1979, Western Mining had the funds it needed to "drill out" Olympic Dam and determine the extent of the resource.

By this time, the size of the discovery — and what it could mean for the recession-hit South Australian economy — were the biggest questions in Adelaide. Just weeks later, Olympic Dam would be a major issue at the 18 September state election. The Labor Government was about to face the polls for the first time in a decade without its popular leader, Don Dunstan. Ill-health forced his resignation unexpectedly in February, with Des Corcoran taking over.

The new Premier went to the election with a half-baked policy; Labor would continue to ban Olympic Dam along with all other uranium mining, but it might change its position if it could be satisfied on safety issues.

The policy was the result of a collision between Dunstan's personal opposition to uranium mining and a growing move in the state Labor Party to allow uranium mining. Dunstan travelled to the UK and Europe just weeks before his retirement to make his own assessment of the safety of nuclear power. He made the trip with his executive assistant, Bruce Guerin, along with press secretary, Mike Rann, the chief of AMDEL (a state government-owned mineral research and testing laboratory), Ron Wilmshurst, and the former director of mines, Sir Ben Dickinson. Wilmshurst and Dickinson were both strong advocates for uranium mining and enrichment in South Australia.

Keith Johns, who was then deputy director general of the Department of Mines and Energy, writes[14] that the trip was generally interpreted as a means of allowing the Government to drop its anti-uranium policy and permit development of Olympic Dam.

"Dunstan returned to Adelaide on 5 February 1979 and declared that uranium would not be mined in South Australia. The decision came as a shock to most

---

[14] *A Mirage in the Desert?* R. Keith Johns, O'Neill Historial and Editorial Services, 2010

observers and it was roundly condemned. WMC expressed 'disappointment' but there was no slackening of activity at Roxby Downs."

In the face of such an obstacle, many companies might have dithered, but Sir Arvi explains that Western Mining pushed ahead in the belief that rational argument would eventually win it approval to develop its great discovery.

"There was never any question of not proceeding with exploration and planning at Olympic Dam. Australia and the Labor Party had been strongly in favour of nuclear energy since the 1950s. Labor, which had now imposed the ban on uranium mining, had imported the anti-uranium stance from overseas and was internally divided on this issue. We did not think a ban could last because it did not make sense.

"We knew that it would take a number of years to get to the stage where we had enough information to even formulate a project, so nothing was held up by the ban on uranium mining. Also, it was necessary to have a definite project proposal to test the politics because this would make it clear what the State would forego in terms of income, development and employment if the ban continued. In the end, this was what made both sides of politics in South Australia become ardent supporters of Olympic Dam."

South Australia's Liberal Opposition, led by David Tonkin, was clear about its stance on uranium mining. It went to the 18 September 1979 election with a promise of development and jobs, including the fastest possible track for Western Mining's huge copper and uranium discovery. The issue developed into an argument about what a ban on the development of Olympic Dam would cost the state in terms of jobs and economic growth. Of course, no-one really knew, but this did not prevent both sides of politics making statements about it. In the final days before polling, the Fraser Government's Federal Minister for Employment and Youth Affairs, Ian Viner, told Federal Parliament that Olympic Dam would create 55,000 jobs (5,000 direct jobs and 10 times this figure in indirect jobs), in addition to 3,000 jobs during construction.

By contrast, Corcoran and some of his key ministers continued to downplay the possible benefits of developing Olympic Dam and its prospects of ever being developed. Soon after taking over from Dunstan, Corcoran had told *The Advertiser*, on 24 March 1979, "there have been exaggerated reports about the economic bonanza that would result from the mining of uranium at Roxby Downs."

On election day, Olympic Dam's promise of jobs and an economic boost for the state exerted a much stronger pull on South Australians than any uneasiness about uranium. The Tonkin government was swept into power by a 10% swing against the Labor Party. In some seats, the swing against the Government was as high as 22%.

On the first business day after Tonkin's win, Prime Minister Malcolm Fraser flew to Adelaide amid a blitz of media coverage. Fraser was reportedly on his way to offer in person "full Federal assistance" to get Olympic Dam underway and other measures to lift the state.

Sir Arvi says Fraser's rushed trip to Adelaide had Western Mining pleased but also worried. "The Federal Government under Prime Minister Malcolm Fraser was fully supportive and urged us on. We became worried that he did not understand that the project was at a very early stage. I briefed him in Melbourne on 20 December 1979, stressing that a large amount of work remained to be done. My concluding comment was: 'The magnitude of the task ... is such that even in ideal circumstances two to three years will be necessary before development plans can be drawn up.'"

Western Mining's exploration director, Roy Woodall, had also been busy hosing down expectations of an imminent bonanza, telling Robert Gottliebsen at *The Australian Financial Review* in the days after the election: "At the time of the joint venture agreement with BP Minerals, it was decided to substantially escalate the rate of drilling of the Olympic Dam prospect. However, it is an enormous job and it is unrealistic to think that any early results are going to come from exploration drilling. We need to be left alone — quietly — for a year before we are asked again where we stand on Olympic Dam."

Western Mining and its new partner BP planned a massive increase in drilling activity. When the joint venture was announced in July 1979, three diamond drill rigs were operating one shift per day across all of Western Mining's prospects in the Stuart Shelf project area. Most of this activity was focused away from Olympic Dam, with only three holes — RD 21, 22 and 23 — drilled in the past year at Roxby Downs. Exploration Division was still responsible for all drilling activity, with John Emerson in charge.

With the formation of the joint venture, Roxby Management Services (RMS) would take over. The new company, wholly owned by Western Mining, was formed to manage the joint venture from its new offices in Greenhill Road, Adelaide. It was a big step up from Dan Evans' garage at Flagstaff Hill, which just four years earlier had been Western Mining's only presence in Adelaide. In the interim, Exploration Division's South Australian office moved from Flagstaff Hill to a house across the road from the Marion Hotel in the southern suburb of Mitchell Park. Jim Lalor, the Melbourne-based leader of the Olympic Dam exploration effort from its earliest days, stayed at the Marion Hotel because it was conveniently located between Adelaide airport and Flagstaff Hill. After striking up a friendship with the owners of the hotel, Lalor thought a vacant house across the road seemed like an obvious choice for an expanded office.

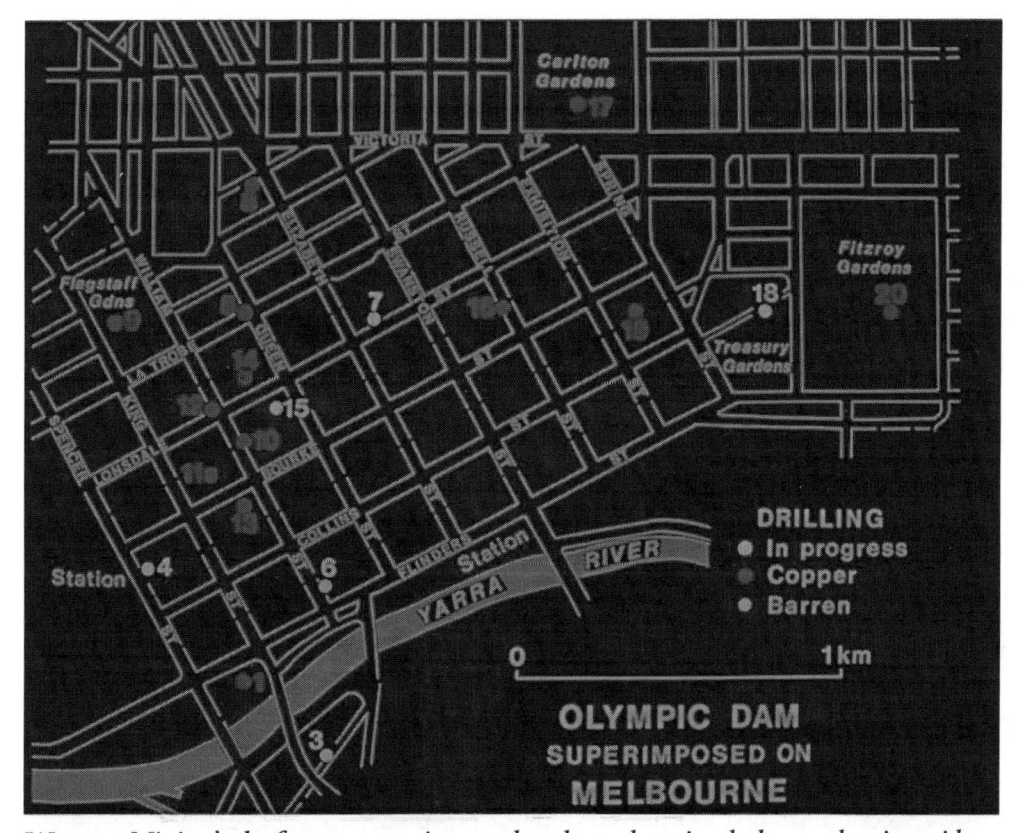

Western Mining's draftsmen superimposed early exploration holes on the city grids of Melbourne and Adelaide to help the company explain the scale of the discovery and the challenge of drilling out the resource. In this map, RD1 is at the western end of Flinders Street. RD20, which helped to finally confirm a discovery, is more than two kilometres away in Fitzroy Gardens.

RMS planned to have 10 diamond drilling rigs operating three shifts a day from the beginning of 1980. This represented a 30-fold increase in drilling activity and would cost in the order of $4 million a month in today's terms. It would also mean a large increase in the number of workers on site. Exploration Division's rigs were manned by about eight workers who lived in three battered caravans near Olympic Dam. Water was carted in once a week from one of the better quality dams near the homestead of Roxby Downs station owner Tom Allison, about 30 kilometres south.

With the onsite workforce now set to rise to at least 60 people, larger and longer term solutions were needed for water supply, sewerage and electricity, along with improved access by road and air. These would be the first of many logistical challenges set by the remote location and hostile climate.

Western Mining knew how to build large camps and even entire towns in the outback. It built the Kambalda township in the 1960s in a physical environment just as harsh as that around Olympic Dam. However, Kambalda was only 50 kilometres southeast of Kalgoorlie, a mining city with more than 20,000 people and supplies of water and power. Olympic Dam was 250 kilometres from the nearest city, Port Augusta, and 500 kilometres from the nearest flowing water on the Murray River.

John Showers, a surveyor who worked for Western Mining at Kambalda and on the redevelopment of the gold mining town of Laverton, was called out of semi-retirement with the challenge of establishing a fully equipped camp, water supply and airstrip at Olympic Dam for 60 workers. A "core farm" where many kilometres of diamond drill core could be geologically logged, sampled for assays and stored was also part of the brief. All of this had to be done in less than four months to allow the joint venture to begin its ambitious drilling program in January 1980.

In a book about his Olympic Dam experiences[15], Showers wrote that the fence line between Roxby Downs Station and its eastern neighbour, Andamooka Station, was also the eastern boundary of the Woomera Restricted Area.

"It was a significant line with the whole of Roxby Downs Station and the area for the proposed intensive drilling being to the west and therefore inside the Restricted Area. We were intent on keeping the camp and the airstrip out of this area to avoid potential delays in obtaining approvals from the remote bureaucracy in Canberra."

Showers and Western Mining geologist, George White, visited Olympic Dam in the first days of October 1979 and selected a site for the 60-man camp on Andamooka Station. They were not pleased with the location — an area of stony gibber inside a horseshoe of low sand dunes — but it was the best they could find outside the Restricted Area and within a few kilometres of Olympic Dam.

Tom Allison had agreed to provide more water from the dams near his homestead to meet the needs of workers who would construct the camp.

Showers wrote that he contacted the Mines Department and the Pastoral Board to obtain approval for the sites selected for the camp and airstrip. "They were sensitive to the location on Andamooka Station as the property was already adversely affected by

---

[15] *Return to Roxby Downs,* John Showers and Bookends Books, 1999

the extensive area of their land taken up by opal mining. We in turn were emphatic that we wished to be clear of the restrictions of the Woomera Restricted Area."

The promises of support from the new State Premier and the Prime Minister then materialised in an entirely unexpected way. "On October 10th Rod Everett from the Pastoral Board contacted us with the news that the Commonwealth had agreed to change the Restricted Area boundary. This would enable us to locate the camp and airstrip on Roxby Downs and would also free up the mineralised area where the intensive drilling was to take place. Apparently, Rod had pursued the matter at a high level resulting in the Premier contacting the Prime Minister and receiving almost immediate approval. This was a very timely and significant decision which enabled the siting of initial and subsequent major infrastructure in the most favourable locations."

Showers, White and Bob Perry from consulting engineering firm Kinhill flew to the site the next day, landing on Lake Blanche, to choose new sites for the camp, airstrip and core farm before the sun went down. A new location was selected about three kilometres south of Olympic Dam. By this stage, construction of a dirt road west from Andamooka towards Olympic Dam had been underway for a number of weeks. The new road would enable Atco to deliver an order of pre-fabricated accommodation buildings, which had recently been increased from 60 to 80 rooms. With only three weeks until delivery, Showers and his growing team of surveyors, engineers, tradesmen and earthmovers would need to be creative problem solvers to meet their New Year deadline.

Earthmoving equipment to build the airstrip was found among the opal fields of Andamooka. The local miners jumped at the chance to earn an hourly rate and quickly assembled a ramshackle convoy of tractors, scrapers, loaders, water tankers and tip trucks. Construction went quickly with the help of Northern Earthmovers, which had now finished building the road from Andamooka to the camp.

Power would be supplied by diesel generators, fueled from tanks installed on site by BP. Telecom had agreed to install six phone lines by setting up a radio link from Andamooka, and even a liquor licence for the canteen would be granted in time. But the biggest worry continued to be water supply. Two potential solutions to the problem had already failed. The first idea was to put a plastic cover on a large dam on Andamooka Station that held more than enough water for the station owner's needs. By stopping evaporation, the joint venture could tap into a supply that would otherwise disappear back into the desert skies. The station owner was unconvinced and knocked back the idea. The second option was to extract and treat the salty groundwater buried deep under the camp. Desalination could make most groundwater fit to drink, but the water in the local aquifers contained a high level of sulphates and could not be used.

Showers wrote he then struck upon the idea of trying an old bushman's trick to make the water in Olympic Dam itself fit for drinking. "We kept looking at the muddy water in Olympic Dam which was of good quality if only we could find a way of cleaning it up. The colloidal clay making it muddy was too fine for filtering so we tried an old bush method on a small sample. Adding a small quantity of salty water was supposed to settle out the mud in the very muddy water. We took samples of salt water from the driller's bore and muddy water from the dam and tried it. It worked. After trying various mixtures we soon found the minimum requirement of salt water to drop out the silt and that the clear water obtained was acceptable for use."

A standard PVC pipe was laid to take the Olympic Dam water three kilometres south to the camp, where it was chlorinated and filtered through a domestic-type sand filter. The small, muddy watering hole known as Olympic Dam had not only loaned its name to a mighty mining venture, it provided drinking water for the first 100 or so workers. This continued until a pipeline from Woomera, 75 kilometres further south could be approved and constructed. Woomera water had already travelled all the way from Morgan on the Murray River via Port Augusta. It was an expensive supply, but it could be counted upon until feasibility studies were completed and the future water needs of the project were known. Everything was set for the new drilling program to begin in January 1980.

By the end of the first quarter of 1980, 10 diamond drilling rigs were on site, along with three percussion rigs to pre-collar holes through the thick sedimentary cover to a depth of about 300 metres. The program consisted of four phases of drilling on progressively tighter grids.

In the first phase, about 35 exploration holes were drilled at the intersections of an 800-by-800 metre grid over an area of about 25 square kilometres. This phase was largely completed by the end of 1980 and succeeded in outlining the boundary between the brecciated rocks that were a crude marker of the zone of interest and the ordinary basement rocks of the Gawler Craton.

This initial phase showed the brecciated rocks were not evenly distributed, with a corridor along a northwest-southeast trend and an irregularly shaped core. This was the first sign the partners would be battling complexity as well as the sheer size of the resource in their efforts to delineate the Olympic Dam orebody.

A follow-up phase of drilling on a 200-by-200 metre grid was well underway by mid-1980. This would be a far more intensive program of about 250 holes over an area of about eight square kilometres. The initial focus was on the eastern side of the resource. The grades were not as high as some other locations, but the consistency of grades across a large area made it an early target for evaluation.

Drilling in the eastern areas progressed rapidly into the third phase of 100-by-100 metres and into the fourth phase of 50-by-50 metres. The results confirmed the consistency of grades. By mid-1981, the partners had developed a detailed plan for a large, underground mine based on the eastern areas, treating 6.5 million tonnes of ore per annum and producing 150,000 tonnes of copper and 3,000 tonnes of uranium oxide.

But this plan changed suddenly late in 1981, despite some 200 holes already drilled in the eastern zones. It would be another three decades before a new mine plan was developed for this area, which today sits in the heart of the giant open-cut operation proposed by BHP Billiton.

The change of focus by Western Mining and BP was triggered by the world recession in the wake of the second oil crisis in 1979. A sharp downturn in global economic activity had already led to a 30% decrease in copper prices in the two years since work began to drill out the resource, and further falls seemed inevitable.

Uranium prices were also in free fall, with the Three Mile Island accident in early 1979 adding to the woes of the nuclear power industry. By contrast, gold prices soared to an all-time high of $US875 an ounce ($US2,200 an ounce in today's terms) as investors rushed to the perceived safety of bullion. But gold prices could never be high enough to compensate for the depressed state of the markets for Olympic Dam's main commodities.

All previous assumptions about the economics of mine development had to be thrown out. Sir Arvi says of this period that, while there was never any doubt in his mind Olympic Dam would be developed, there was a real prospect of the project being halted for some years in the same way development was suspended by Alcoa at Portland.

"The early 1980s, following the second oil crisis, were a very difficult time. At Alcoa, a brand new alumina refinery just completed at Wagerup in Western Australia was mothballed because of a serious downturn in markets. In Victoria the building of the Portland smelter, started in 1979, was stopped half completed in 1982 because the dreadful market was combined with an announced 25% increase in the cost of electricity. This made the smelter uneconomic and to complete it we had to borrow hundreds of millions of dollars, which was real money at that time. Apart from our own unwillingness to put the company in jeopardy, the lenders were not happy to advance further funds in these conditions.

"Work was not resumed until 1984 after markets had picked up, the electricity tariff had been renegotiated, and Alcoa had acquired partners in the smelter able and willing to take the long term view, including the Victorian Government. The Wagerup refinery was also started up in 1984."

He says everyone believed markets would recover, but no-one could predict when. "We had to find a way to justify continuing the work at Olympic Dam, even if the dreadful market continued for a long time. Finding higher grade ore was the answer, and we were lucky to have such areas. Had we not had these, we may have had to stop the project for a while until conditions improved, as we had to do at Portland."

The Olympic Dam partners suddenly needed higher grades, and the answer seemed to be coming from drilling in the narrow corridor of brecciated rocks in the northwest. A number of drill holes on the 200-metre grid had recently assayed copper grades of 4% and even 5% copper, compared with the average 2% grade in the eastern areas. Uranium oxide grades were also two or even three times better. Drilling activity was refocused on the northwest in the hope of establishing enough high-grade reserves for a smaller underground mine. The new plan had many advantages; a smaller, high grade mine should be economically viable even if prices continued to fall, the costs to build the mine would be lower and the development time shortened.

There was still a hurdle to overcome with the new plan. It relied on Western Mining's ability to prove sufficient high-grade ore within complex mineralisation at depths of half a kilometre and more. This would be difficult for any miner, even with a fleet of drill rigs at its disposal. The challenge was even greater at Olympic Dam because there was no precedent for this type of orebody. No-one knew how it was formed. If there were any patterns in the rock units and mineralisation to guide the drilling program, no-one knew what they were. Patient and methodical drilling over hundreds of thousands of metres lay ahead.

In a 1988 paper for Australasian Institute of Mining and Metallurgy, Western Mining's chief geologist at Olympic Dam, Jim Reeve, wrote that the high-grade, D Areas were deceptively complex. "The 100 metre grid holes produced some spectacular intersections, but they also indicated that this area was much more complex than (the eastern zone). Further in-fill drilling to a density of 35 metres by 70 metres spacing was completed over much of Area D, and eventually to a spacing of 35 metres by 35 metres in Area DNW in an attempt to resolve the geological complexity of the most strongly mineralised area. However, even at this very close spacing there was insufficient information to allow confident interpretation of ore zones or lithological units. Most interpretations from this period were later shown to be incorrect."

For the time being, however, Western Mining had confidence Area D could support a small, high-grade mine. It believed it had found a way to continue development, despite a world recession and a highly uncertain future for commodity prices.

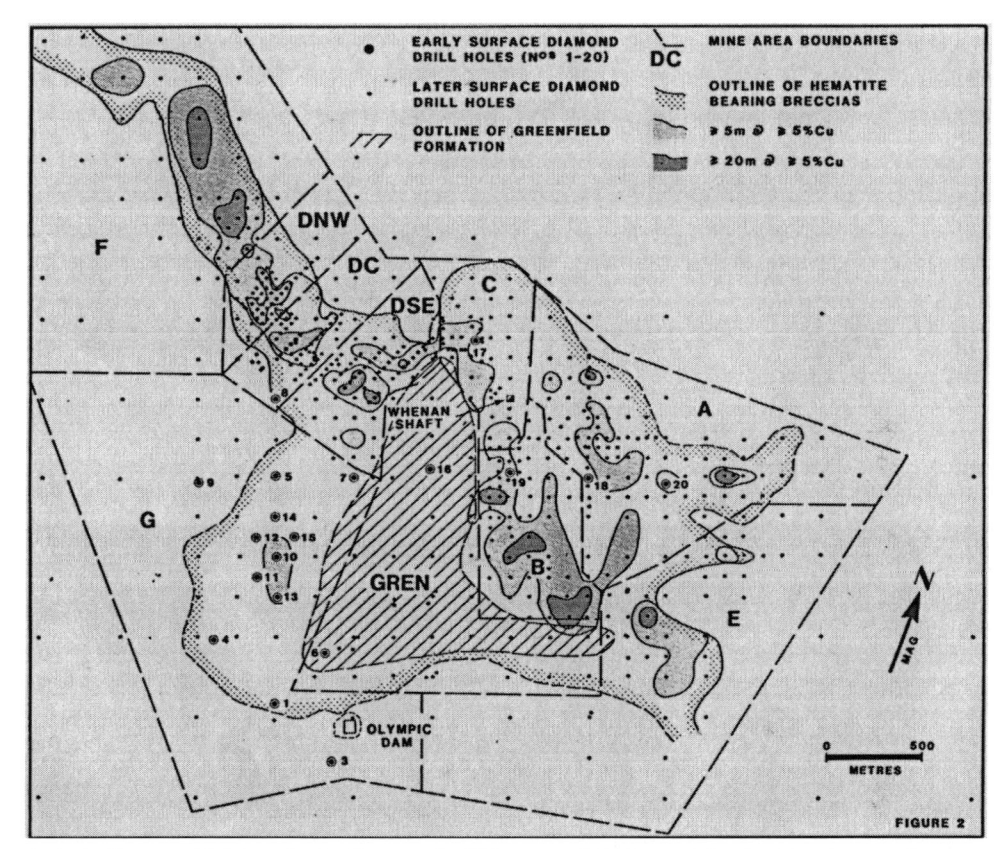

A map from a 1988 technical paper by Western Mining geologist Jim Reeve shows Olympic Dam divided into areas (A to G) to simplify the massive task of delineating the resource. Areas A and B in the east were the initial target, but a global recession in the early 1980s forced a change in focus to high-grade mineralisation in Area D (DNW, DC, DSE) in the northwest.

Buoyed by this achievement, the joint venture partners finally released their first estimate of the size of Olympic Dam. It was almost two years since the size of the resource became a major issue in the South Australian election, but public interest had not cooled. The company announced on 26 July 1982 that Olympic Dam contained an estimated resource of 2,000 million tonnes, averaging 1.6% copper, 0.6 kgs/tonne uranium oxide, 0.6 grams/tonne gold and 3.5 grams/tonne silver.

News media around Australia reported that Olympic Dam ranked as one of the biggest mineral discoveries in the world, with some of the most excited coverage by Adelaide's afternoon tabloid, *The News*. "Roxby Downs is BIG," wrote Brian Francis.

"It's bigger than we imagined and it will be pumping out ore when other mining ventures breaking news today are long forgotten. Today, financial and mining circles are a-buzz across Australia, across the world."

The article quoted Western Mining's executive director of operations, Keith Parry, about the progress of drilling. "We are terribly excited.... It is a magnificent deposit of ore — copper, uranium and gold — which starts about 350 metres below the surface ... It extends over an area of about 7 kilometres by 4 kilometres. We have had 13 drills working for two years in this zone. We are closing in now on significant tonnages of high grade ore in two areas, roughly about half a kilometre by half a kilometre within the larger zone."

While Western Mining's understanding of the high grade ore zones was still far from perfect, the task of drilling out the resource would soon become easier with the completion of a 500-metre deep exploration shaft in September 1982. It was named the Whenan Shaft, in line with Western Mining practice of naming mineshafts after the driller who completed the discovery hole. Roy Woodall began the tradition as the company's way of paying respect to the tough drillers without whom discoveries would not be made, no matter how brilliant the company's geoscientists.

The partners began sinking the 6.5 metre-by-3.5 metre shaft in mid-1980 at a cost of $15 million (almost $45 million in today terms). This was in addition to the $50 million ($195 million) already committed to explore and evaluate Olympic Dam when the joint venture was formed in 1979. The shaft was added to an already massive program of activity after Western Mining became convinced that surface drilling was not going to provide the certainty it needed. The company had grown up around the notoriously difficult underground gold mines of Kalgoorlie, where experience had taught the necessity of getting underground as early as possible. It was a wise call, made all the more important by the switch in target from the large, medium grade areas in the east to the small and complex high grade ore zones in the northwest.

The Whenan Shaft was sunk inside the eastern zone, reflecting the thinking in mid-1980 about where ore for a future mine would be sourced. Area D was between one and two kilometres northwest of the shaft, but a horizontal tunnel or drive into this high-grade zone would now be the top priority. Underground exploration drilling from the drive into the northwestern areas soon became the primary means of evaluating the high-grade ore zone. Drill holes at spacings of just 17.5 metres, fanning out along an increasing number of horizontal drives, would finally provide the certainty Western Mining demanded in its ore reserve calculations.

The other main objective of the Whenan Shaft was to allow the extraction of large samples of potential ore for metallurgical testing. The highly complex and uncertain task of developing a process to extract copper, gold and silver from Olympic Dam's incredible mixture of uranium minerals was about to begin in earnest.

By mid-1987, the joint venturers had drilled from the surface a total of 234,000 metres of diamond drill holes and 240,000 metres of percussion drill holes. Drilling from underground now also totaled a massive 143,000 metres.

The first statement about probable ore reserves was released in November 1983, about a year after about underground drilling began. The joint venturers announced a probable ore reserve of 450 million tonnes, averaging 2.5% copper, 0.8 kgs/tonne uranium oxide, 0.5 grams/tonne gold and 6 grams/tonne silver. The average copper grade in this probable ore zone was markedly improved from the 1.6% average announced over a year earlier for the entire resource, and highlighted the progress that had been made in defining the basis of a high-grade mine. While 450 million tonnes was a small component of the total resource, it was still enough to make Olympic Dam bigger than the reserves behind Australia's largest existing copper mine at Mt Isa in northwest Queensland.

⚒ 14 ⚒

POLITICS AND PROTESTS

With the Liberal Party under David Tonkin elected in September 1979, the joint venturers now had State and Federal Governments in favour of Olympic Dam. Within weeks, work began on a detailed agreement setting out the rights and obligations of the joint venturers and the State Government in the development of the resource.

It was called the Indenture Agreement and was intended to go before the South Australian Parliament to be voted into law. For Western Mining and BP, the Indenture was all about creating a safe foundation on which to base a decision to invest several billion dollars over the next decade. These types of agreement are normal for major resource projects, which are risky enough without the possibility of a new state government changing the ground rules. Western Mining had negotiated similar agreements for its nickel operations at Kwinana and Kalgoorlie.

Getting the Indenture through the House of Assembly or Lower House would be straightforward. The Government held the majority of seats and would vote the Indenture through as long as the joint venturers had met all of the Government's concerns in framing the agreement. However, the Legislative Council or Upper House, equivalent to the Senate in Federal Parliament, loomed as a major hurdle. The 21 seats in the Upper House were evenly divided between Liberal and Labor, with one Australian Democrat, Lance Milne, holding a deciding vote.

The Labor Party was still opposed to developing Olympic Dam, despite this position playing a major part in it being dumped from office at the 1979 election. In June 1981, the annual state convention of the Labor Party called for South Australia to be declared a nuclear-free zone. Three months later, in September 1981, the national conference of the Australian Democrats resolved that Olympic Dam could proceed, provided the uranium was put back in the ground.

Sir Arvi says the politics associated with uranium mining was the biggest problem faced by the joint venturers. "We knew how to drill out and mine orebodies and we could figure out metallurgical processes, even for a unique and complex resource such as Olympic Dam. These were technical challenges we understood and over which we had control, but the politics of uranium mining was completely outside our experience."

Western Mining's response to opposition from the Labor Party and the Democrats was to keep pushing ahead on all fronts and persist with rational argument for developing Olympic Dam. This approach had worked in 1979 when the company found a joint venture partner to get on with drilling out the resource, despite the state Government ban on uranium mining. Sir Arvi was determined to keep going, and drew encouragement from signals he received from sources away from politics and the public protests of an increasingly noisy minority. "It was significant that while left-wing union leaders were anti-uranium in public, their members continued to work on the project. The main minerals industry union (AWU) was, if anything, supportive. This, and the public opinion in South Australia, were both important indications that the opposition to Olympic Dam could not last."

As an organisation, Western Mining could not have been more removed from the stereotype of a bloated mining company with an army of political lobbyists. In the early 1980s, the company still did not have a public relations department. Responsibility for communicating the company's views to politicians and community groups was assigned to one of its metallurgists, John Reynolds. He had been appointed to the new role of Manager, Corporate Affairs in about 1977, when protests in Western Australia over the company's Yeelirrie uranium project were growing louder.

In his unpublished memoirs, Reynolds writes that from the mid 1970s, Western Mining and others felt it was necessary to put more effort into countering the public campaigns being mounted by opponents of uranium mining.

"The Australian Uranium Producers Forum was established to provide information and to reply to the often subjective and ill-informed criticism being aimed at the industry and demands to stop its establishment. We also felt the need for a communication channel to the national and state governments and parliamentarians. As part of this work industry people accepted invitations from various organisations to participate in debates or give talks on aspects of the nuclear industry from mining through to power generation and the international issues of nuclear disarmament. Our "opponents" included some articulate and dedicated people who were only too willing to appear for the other side of the argument. We prepared ourselves with study and the assembly of relevant information. In the early stages I was drawn into the exercise in Melbourne, mainly to participate in debates arranged by branches of the Australian

Labor Party. I recall the first one at the Highett Branch, which as in the early TV exposures, was somewhat frightening at first, because the audience was noticeably hostile and regarded anyone who was in favour of uranium as being immoral and a villain of deepest dye. However I got through that and realised that my opponents were often light on facts and tended to be high on emotion. Some had done their own homework, but drew conclusions which were either wrong or lacked objectivity regarding risk assessment. The second encounter was, I think, at Frankston and that was not so frightening, although I was still a villain. After that I started to enjoy the many subsequent exercises around suburban Melbourne and occasionally in the country, as a sort of verbal game in defence of something which I believed was a worthwhile cause to support. On one occasion at the Nunawading municipal centre an academic opponent made a strong personal attack on me from the platform, implying that anyone concerned with uranium was verging on being criminal. This was not pleasant but part of the job. Politicians no doubt regard this sort of thing as a normal occupational hazard.

"In the further debates around the Melbourne suburbs, I noted that a group of the same people were usually in the audience, whether in Essendon or Nunawading. We almost struck up a mutually tolerant acquaintanceship and I grudgingly admired their persistence. They would ask the same questions, often at length, at each meeting and I would provide the same reply for the benefit of new members of the audience. But it seemed to me that the regular attendees were quite uninterested in the answers, and the important thing was the asking of the question. Over time this trend caused us to change policy and refuse invitations to debate, but offer to speak to any audience and to answer questions. A consequence was improved communication and information became somewhat more important than 'who won the debate.'"

One of Reynolds' responsibilities was to prepare Western Mining's submission to a Select Committee of the Legislative Council, set up by Tonkin after his election in 1979, to investigate all aspects of mining uranium in South Australia and in particular the safety of workers. The Committee comprised three Labor members, three Liberal and the Australian Democrat, Milne. One of the Labor committee members was Norm Foster, a former organiser of the waterfront union, who would later become a key figure in the approval of the Indenture Agreement.

The Indenture Agreement was finalised with the Tonkin Government in March 1982 after two years of intensive negotiations between Hugh Morgan and his team for the joint venturers and the Minister for Mines, Roger Goldsworthy. On 9 June 1982, the Indenture Bill was passed in the Lower House. Two days later, *The Australian* ran a headline 'All hope lost for Roxby' on the basis that unyielding opposition from Labor and the Democrats in the upper house would sink the bill. But Labor's opposition to

Olympic Dam was weakening. On 13 June, the annual convention defeated a motion to declare the state a nuclear-free zone by 92 votes to 68. In his personal notes about these events, Sir Arvi writes: "Norm Foster, who had been having further telephone discussions on radiation and safety with John Reynolds, told the convention that he would be under great trauma and mental strain to cast a vote against the indenture. He said the Party was handing the Liberals an election victory 'on a plate' and he did not want to see the Party self-destruct. ' ... those who are not committed to a political line ... are seeing the facts of mining and milling of it (uranium) are not so dangerous as we would lead them to believe.'"

Indenture key points

The original Indenture agreement covered the construction and development of Olympic Dam up to a production level of 150,000 tonnes of copper per annum. The agreement has been modified over the years with the expansion of the mine, but the key elements are unchanged.

- The joint venturers were to pay a royalty of 2.5% in the first five years, rising to 3.5% thereafter on an ad valorem basis. This means the royalty is based on the value of sales from the mine. The Indenture included an additional 'windfall' royalty, although this was not triggered in Western Mining's time.
- South Australia was to provide schools, a hospital, a police station and recreation and sports facilities in the new town of Roxby Downs to house the mine's workers, as well as carry half the cost of a road from Pimba to Olympic Dam. The joint venturers were to meet the costs of supplying water, sewerage, electricity and all other roads. The Indenture provided for spending of $50 million (in 1982 dollars) by South Australia on infrastructure, based on an expected town population of 9,000. In fact, the number of residents at Roxby Downs settled at about 5,000 and government spending on infrastructure was reduced from the planned level.
- The joint venturers were required to obtain labour, materials and services within South Australia as far as reasonably and economically practical. A similar provision applies to processing of the mine's output in South Australia, which has been a source of tension between the state and BHP Billiton in the proposed expansion plan.
- The joint venturers were required to submit plans every three years for protection and management of the environment throughout its life, over and above the commitments given in the original Environmental Impact Statement for the development.

"At 2 a.m. on 17 June the Legislative Council voted on the Indenture Bill, which was defeated by eleven votes to ten. Norm Foster voted against the Bill. At 9 a.m., Foster sent a telegram to the State Secretary of the ALP, saying he had resigned from the Labor party. At 10 a.m., Premier Tonkin called for a special meeting of the Legislative Council to vote again on the Indenture Bill. In a special meeting on 18 June, Norm Foster crossed the floor to vote with the Liberal Government and the Bill was passed eleven votes to 10. (Opposition Leader) John Bannon announced that an ALP Government would amend the Indenture Act only 'with the co-operation of the Joint Venturers.'"

One of Foster's Labor colleagues in the Upper House, John Cornwall, later wrote in his memoirs the Party had hatched a "clever and cruel" plan to goad Foster into crossing the floor. This would allow it to achieve its real objective of having the Indenture passed, without being seen to change its position.

Reynolds, who was in the public gallery at the time, writes that he recalls the jeers that were poured onto Foster and admired his courage. "There has been speculation about the motivation for these events but I think Norm genuinely changed his views due to certain events outside parliament of a personal nature which steeled his resolve, together with the information with which he had been provided and exposure to situations overseas where uranium mining and the nuclear electricity industry was accepted."

Foster's decision to cross the floor and Labor's unwillingness to challenge the Indenture effectively removed Olympic Dam as an issue at the upcoming state election on 6 November 1982. A few weeks before the election, Bannon announced that Labor would not stop development of Olympic Dam. He was voted into office and became one of the state's youngest premiers at the age of 39.

Four months later in the March 1983 Federal election, Australians replaced the Liberal-National Party Government of Malcolm Fraser with a Labor Government under its new leader, Bob Hawke. Political commentators speculated the election result could be a fresh blow to Olympic Dam, but the Federal Labor Party had already started to unwind the anti-uranium policy adopted in 1977. At its July 1982 conference in Hobart, the Federal Labor Party amended its platform to allow production from any new uranium mines and to consider mining and export of uranium incidental to mining of other minerals. It was a nonsensical policy that effectively deemed some mines produced bad uranium and others produced good uranium, but it gave Federal Labor room to approve Olympic Dam and thereby avoid a voter backlash at the upcoming Federal election.

Morgan says Hawke and Bannon led the push for the policy change at the 1982 conference that would become known as the Three Mines Policy. "Hawke and Bannon took on the left-wing of the party and they put their political lives on the line. Bannon's commitment was remarkable. He had become a believer in Olympic Dam and was critical in bringing about the change in Labor policy."

Development headaches

While the politics did not slow the development of Olympic Dam, the emotionally charged issue of uranium mining and the public protests that went with it had a major influence on the way Olympic Dam was developed. According to Sir Arvi, the fact that development approval was narrowly won made Western Mining and BP even more wary of the risks of developing Olympic Dam.

The partners already had technical and market reasons to develop the project with the lowest possible investment. The political risk of approval being withdrawn by a future government was an additional reason to take this approach. "We didn't want to risk more than the minimum necessary investment, so we opted to develop it as an underground mine. Even a small open cut would have cost many additional hundreds of millions to remove the overburden."

He says mining underground had the additional benefit of making Olympic Dam a smaller target for uranium protestors. "The anti-uranium campaigners would do anything to stop the project, and going underground would deprive them of the additional arguments against the higher visual impact of an open-cut. The main reason, however, for underground mining was the ability to develop the mine as we went and having less capital at risk up front."

The anti-uranium climate of the 1970s was also a factor in Western Mining's decision to build a bigger and more costly processing complex at Olympic Dam. Most copper mines do not process ore beyond the stage of a sulphide concentrate, which contains anywhere between 35% and 75% copper metal, depending on the type of copper sulphide being mined. At this stage it becomes economically sensible for most mines to ship the concentrate to large smelter/refinery complexes in places such as Japan and Germany, where it is treated on a fee-by-the-tonne basis. Smelting turns the concentrate into blister copper with about 99% purity. The further step of refining turns blister copper into copper cathodes with a purity of at least 99.99%.

In the treatment process planned for Olympic Dam ore, uranium oxide would be leached from the concentrate before it was sent to the smelting stage. In theory, this would make the concentrate suitable for sending offsite for custom smelting and refining. But trial processing of the ore in a pilot plant at Olympic Dam revealed the leached concentrate still contained residual uranium and uranium decay products. Shipping this to custom smelters would not be possible because it might contaminate copper products smelted for other companies.

Western Mining's metallurgists advised that the company would need to build its own smelter to treat Olympic Dam concentrate on site. A batch was sent for smelting trials to Finnish company, Outokumpu, which had developed an innovative, one-stage process. The blister copper from the smelter trials — now about 99% pure — was finally clear of uranium, but it still contained radioactive traces of uranium's decay products such as lead and polonium.

Morgan says the company then looked closely at whether it would need to also build a refinery to process Olympic Dam copper on site to the very end stage of copper cathodes. "We realised at quite an early stage that we might need to build a smelter and a refinery at Olympic Dam because of the uranium content of the ore. This not only meant a heavy capital burden, but it was very risky from a commercial point of view to build these types of facilities in the middle of the South Australian outback. These would be stranded facilities, a long way from any shipping port. If for some technical reason we couldn't process the Olympic Dam ore, we were too distant to treat copper concentrates from other producers as a back-up plan. So you wouldn't build a smelter/refinery complex at Olympic Dam unless you had a special reason.

The metallurgist's paradise

Sir Arvi says Olympic Dam is "a metallurgist's paradise" and developing processes for extracting copper, uranium, gold and other economic minerals was the second biggest challenge after the politics of the mine. "The metallurgical treatment of the ore is not complicated, but it's extensive. We had to go through a very extensive process and be satisfied there were no bugs in it." All of the processes were proven elsewhere but no-one knew whether the unique combination of steps required at Olympic Dam would create unexpected problems. The joint venturers proceeded cautiously, beginning with studies in Adelaide in 1980 by a group of four metallurgists, led by Henry Muller. The team designed a pilot plant using bulk samples extracted via the new Whenan Shaft. The pilot plant was built on site in 1984 with a capacity to treat 45 tonnes of ore per hour. It produced a 500 tonne parcel of copper concentrate for testing in Finland. The pilot plant also included a test of the all-important hydrometallurgical process to extract uranium from the residue or tailings of the copper concentrator. Most of the uranium would be extracted by leaching the tailings with sulphuric acid, which would be made on site from sulphur contained in the smelter gases. Testing confirmed the metallurgical circuit would work and it remains largely unchanged.

"That special reason soon evidenced itself. At BP's request, we began environmental investigations of suburbs surrounding some of the major custom copper smelters. That study found evidence, albeit quite minor, of radioactivity. When we looked further, we found that it was not unusual for copper concentrates from mines all around the world to contain minor amounts of uranium. If we sent Olympic Dam ore for treatment at a custom smelter where there was already radioactivity in the surrounding environment, we would get the blame, even if we were satisfied we had already leached out all the uranium. BP was especially sensitive to this — it was worried that every BP station in the world would be boycotted! We decided that all our production had to be squeaky clean, and so the decision was made to build a smelter and a refinery."

The processing plant designed for Olympic Dam was already extensive and this overshadowed the significance of adding a refinery to the concentrator, hydrometallurgical plant and smelter already planned. But the refinery added hundreds of millions of dollars to the cost of developing Olympic Dam and created further demands for materials, electricity and water in the middle of the desert.

Protests

Protests against Olympic peaked in 1983 and 1984, when blockades of the mine site were organised by a group called Campaign Against Nuclear Energy (CANE). The aim was to stop work and pressure the government to adopt anti-uranium mining policies. Reynolds says the protestors stated various concerns, including claims that radiation from the mine would cause genetic mutations, and that uranium from Olympic Dam would be used to build nuclear weapons. "There was a lot of uninformed nonsense and very little effort to understand the facts. At one time, an academic from overseas told a meeting the entire population of Adelaide was in danger of genetic mutations."

The 1983 protest involved about 400 demonstrators. The joint venturers at that stage had only an exploration licence on a pastoral lease, which allowed entry by anyone. Reynolds writes: "This resulted in several skirmishes as the protestors tried to stop the mine crew accessing the shaft area, usually when the afternoon shift was arriving ...... Some of the protestors did strange things. One super-glued his hands to the mine fence so he couldn't be dragged off. Our engineers threw a bag over his head and cut off a section of the fence, sending the gentleman off holding on very tightly to two short lengths of galvanized steel tubing."

The 1983 protest failed to cause any delay or interruption, so CANE mounted a bigger and more organised protest the following year. In August 1984, 700 protestors descended on the site, but the joint venturers were also better organised. Construction work was now sufficiently advanced to secure a Miscellaneous Purposes Lease under the South Australian Mines Act, allowing the joint venturers to refuse entry at their discretion. The site was fenced and barricaded by South Australian police. At one stage, protestors invaded the site, leading to 300 arrests, but the main protest dissipated after nine days, again without interruption to the development.

Most of the protest activity ended at this time, but Morgan says serious threats to the water supply to Olympic Dam continued well into the 1990s.

"Some of the protest action was particularly vicious. A small group set out to disable the pumping stations on the pipeline that took water from our borefields on the Great Artesian Basin to our desalination plant near the mine site. If they had wrecked the pumps there would have been a long lead-time to replace them, and their objective was to denude or deny water to the operation to such an extent we would have to send people off. The desalination plant was providing about three megalitres a day into storage dams, which was clean water for the town as well as process water for the operation. These pumping houses had barbed wire and electric wire around them and there was even a guard inside with a dog. Well, the protestors drove a car through all of this, they beat up the guy and tried to wreck the pump. But the pump had over the top of it a piece of heavy steel protection and there was intervention to stop them before they could get in. As a consequence of that experience, it was resolved that each of the valves along the pipeline needed to have a concrete cover put over the top of it, and this cost a great deal of money!" Morgan says heavy protection of the pipeline continued well into the 1990s. "At one stage we had a plane flying up and down the pipeline with an observer. The locals were good at telling us if protestors had arrived because these people would stay at local pubs and we would quickly be informed if a problem was looming."

POLITICS AND PROTESTS

15

Epilogue

History shows that all of the obstacles — financial, political and technical — were eventually overcome. Olympic Dam stands today as Australia's largest copper mine and largest underground mine.

History also tells us how difficult it was to build. Ten years passed between the confirmation of a major discovery and the 1988 opening ceremony, even in the hands of a company that did not wait for governments to change their position on uranium mining. Another 12 years passed before the mine and processing plant reached production levels close to those originally intended.

The initial annual production levels of 45,000 tonnes of copper and 1,700 tonnes of uranium oxide were barely economic, but Western Mining and its new partner were being conservative and patient.

Sir Arvi says: "We knew this wasn't an economic level. But we decided we really wanted to be absolutely confident there were no operating problems in the system, in the mining, the treatment and in the production of products that would be marketable.

"And also we wanted to be able to make sure there were customers who would take the product. This was a kind of pilot operation. We didn't lose any money on it, but we also didn't make any money on it. It was 1.5 million tonnes of ore a year, which turned out to be 45,000 tonnes of copper. We thought this was a big enough scale to find out if there were any problems, and on the other hand it wasn't a crippling investment. If it turned out for some reason or other not being viable, then we could have all lived with it.

"But that was always the first stage to prove the project in a market sense, an operating sense, and from then on we would expand it as we were able to do so. And the market was very important. We had to be sure there were customers for the product, both for the uranium and the copper. The important thing was that the copper couldn't have any radioactivity in it. It had to be absolutely pure, and that's why we built the refinery on the site."

Annual copper production grew in a series of steps — increasing to 66,000 tonnes in 1992, then 85,000 tonnes in 1996 and finally 200,000 tonnes in a major expansion between 1996 and 2000.

The sums invested in Olympic Dam are another measure of the difficulty of bringing it to life. Exploration, feasibility studies and a pilot plant to test metallurgical processes cost almost $450 million in current dollar terms. This was just to get Western Mining and BP to a point where they could decide whether to go ahead with the mine.

The original development was budgeted to cost $1,200 million, with another $200 million spent on each of the expansions in 1992 and 1996. The major expansion between 1996 and 2000 cost a further $2 billion, lifting the total capital investment to about $4 billion.

Getting the mine to this stage was still not a licence to print money. In fact, Hugh Morgan speculates that Western Mining put more cash into Olympic Dam in real terms than it took out. "It suffered from almost two decades of lousy prices. Uranium was a particularly bad performer. In the original feasibility study, it was assumed the uranium price would be around $US30 a pound. In fact, the price kept falling to under $US10 a

The development story still to be told

The Olympic Dam Story describes the main challenges overcome in developing Western Mining's copper-uranium-gold discovery, but it is beyond the scope of this book to provide a full account of 13-year development history. There would not be an Olympic Dam without the key people who designed, tested and developed the mine, the processing plant, the township and services of Roxby Downs, and who marketed Olympic Dam products around the world. Keith Parry, WMC's director of operations, was in overall charge of all WMC operations, including the Olympic Dam development, until May 1986. John Oliver was involved as general manager (projects) until 1981. Henry Muller was appointed chief metallurgist in 1980, with responsibility for metallurgical evaluation, pilot plant testing, and the design of the processing flowsheet. Responsibility for managing the Olympic Dam project was carried from January 1980 successively by John Copping, Tony Palmer, Graeme Sauer, Ian Duncan, Pearce Bowman and Peter Johnston. At the mine site, management from June 1984 onwards was successively the responsibility of Robert Crew and Ian Lawrence. Ian Smith and Trevor Peters then shared the responsibility for mining and processing respectively until Ian Smith became general manager in 2000.

pound. While this was a spot market price and less than we received under long-term sales contracts, the poor uranium market really affected the returns from Olympic Dam. We had 15 years of drought in uranium, and copper was pretty average."

Copper always produced most of the revenue from Olympic Dam, but the price of uranium had a major influence on the mine's total return. If the price was good, uranium was a thick icing on the cake. If prices were low, uranium could swing the entire mine from profits to a break-even result, or even worse if copper prices were also depressed.

The accompanying charts of copper and uranium prices show that Western Mining's timing with Olympic Dam was incredibly unfortunate. Price trends prevented it reaping the real rewards of the project, despite patient investment in exploration and development over almost half a century from 1957 until its takeover by BHP Billiton in 2005. Chart 1 shows average annual US dollar prices for copper on the London Metals Exchange since 1950 in both nominal terms (i.e. the dollar values of the day) and in current terms (i.e. adjusted for US dollar inflation). In real terms, average annual copper prices peaked in 1966 at $US10,300/tonne or $US4.66/pound. This was even higher than the peak price in the recent resources boom that led to death-defying thefts of copper cable from metropolitan rail networks. This perspective makes it easy to understand Sir Lindesay's enthusiasm for taking Western Mining into copper in the 1950s and 1960s.

By the time of Olympic Dam's discovery in 1976, the real price had almost halved to $US5,400/tonne or $US2.44/pound. When construction of the mine began in 1986, prices had halved again to $US2,800/tonne or $US1.28/pound. Prices improved over the next four years, but then began a 14-year downward trend under the weight of large, low-cost supplies from new open-cut mines in Chile. The new mines capitalised on the development of more efficient methods for mining and transporting huge volumes of overburden and ore, as well as new copper leaching technology. Copper production costs sunk to historic lows and prices inevitably followed in a global market of weak demand. In the last five years of the long downward trend (2000 and 2004), the price averaged only $US2,300/tonne or $US1.05/pound and dropped as low as $US1,900/tonne or $US0.86/pound in 2002.

Unhappily for Western Mining, the worst price period for copper in at least 60 years arrived immediately after it doubled its investment in Olympic Dam to boost production to its current capacity.

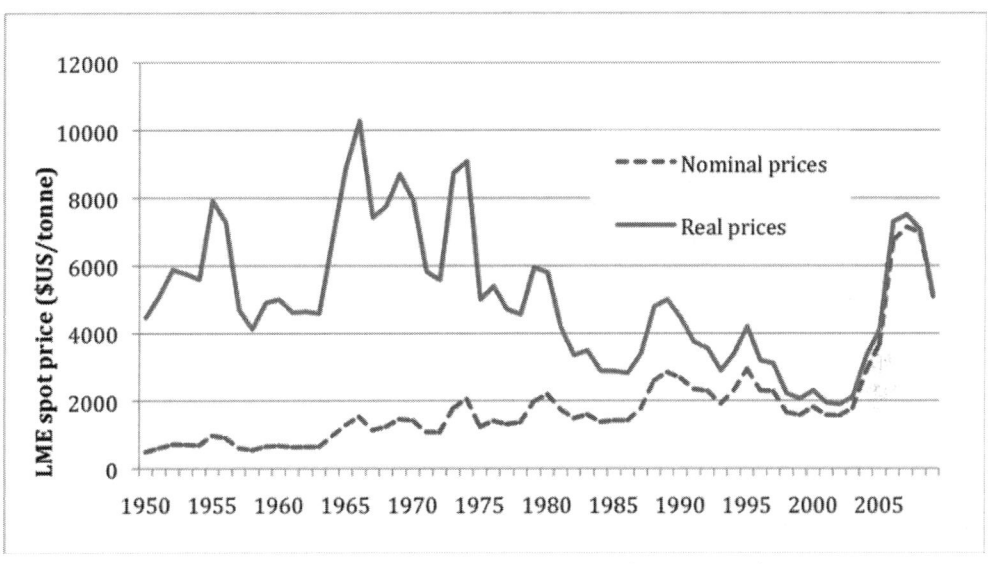

Chart 1: Six decades of copper prices

Source: ABARE Australian Commodity Statistics Data 2009

Actual prices after the expansion were much lower than the price of $US1.40/pound[16] that had been assumed, even though Western Mining had been pessimistic in choosing a figure 25% less than the previous year's actual price, and barely above the lowest prices of the past decade.

Chart 2 shows Western Mining was also unlucky with uranium prices. When Olympic Dam was discovered in 1976, prices were soaring as energy utilities around the world scrambled to lock in supplies for a planned boom in the construction of nuclear power plants.

The Three Mile Island accident in the US in 1978 changed public acceptance of nuclear power. Many of the planned new reactors were never constructed and uranium prices crashed to $US30/pound in real terms by the mid-1980s. In 1986, the market showed signs of turning around, but this time the Chernobyl accident devastated the outlook for the nuclear power industry and sent uranium prices into a deep freeze for two decades.

---

[16] Western Mining board papers. A long-term price forecast of $US1.00/pound in 1996 dollars was assumed when the expansion was approved. This is equivalent to $US1.40 in current terms.

In a cruel twist for Western Mining, the long drought in copper and uranium prices ended just as the company was taken over by BHP Billiton in 2005. Prices finally began to improve in 2003 and 2004 after almost two decades of decline. This would prove to be the beginning of a China-led resources boom, but Western Mining was no longer around to reap the rewards for which it had worked so hard.

Sir Arvi says Olympic Dam was still in a developmental stage in 2005 and would have been expanded again by Western Mining if it had the opportunity. "If WMC had remained in charge of this, there would have been a further expansion. I don't think we would have gone up in such a big step as BHP is planning. I think we would have expanded the underground mine and gone from something like 200,000 to 350,000 tonnes, not 700,000. BHP doesn't have any limitations on capital and of course the market has changed. In all my time there was always the question of the market — can you really write long-term contracts for the output of uranium? The copper was not a problem once the purity was known to be high. You will always sell copper metal, except perhaps in times of extreme economic downturns, but we never got to the stage where this would have worried us. The real question was the yellowcake."

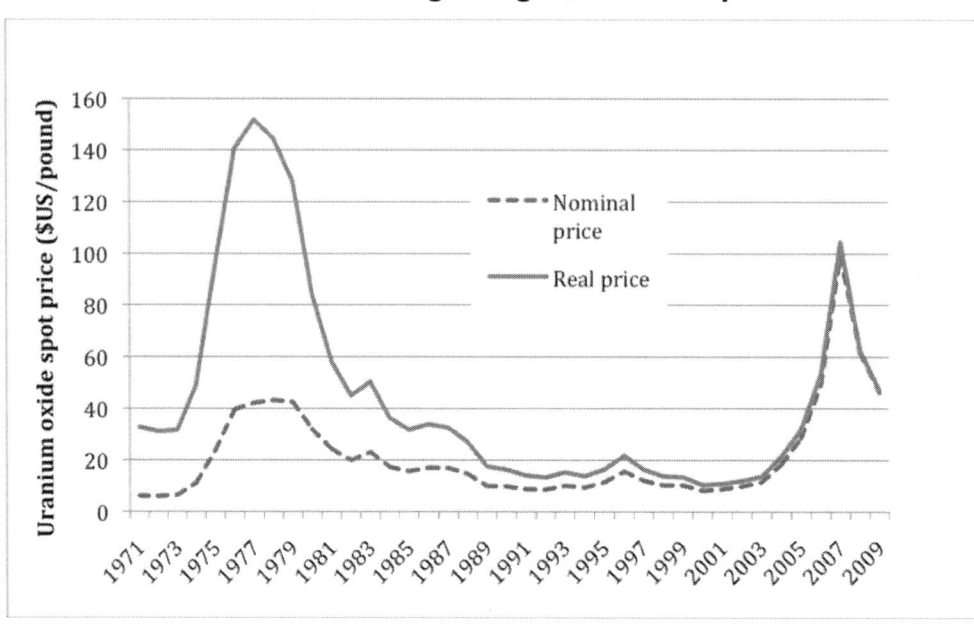

Chart 2: The long drought in uranium prices

Source: ABARE Australian Commodity Statistics Data 2009

EPILOGUE

He says people today may not appreciate the world is very different for resource companies because of China's growth as a minerals consumer. "China didn't really hit its straps as a minerals consumer until 2003. (That trend) is only about six or seven years old. We seem to think it has been there forever, but it hasn't. The whole scenario for Olympic Dam now is very different to what it was until then."

He says in the second half of the 1970s, Western Mining's options for developing Olympic Dam were "quite limited in a number of ways. But we weren't unhappy about it. We thought it was sensible to go step-by-step and make sure we were confident to go to the next step. We realised we weren't going to make much money out of it for a while, but this was a process of long-term development. We had to accept this."

"There was no problem with it at the time because nobody else would have done anything different. If BHP had got in there at the time they wouldn't have done anything different. I mean, we had BP as our partner, one of the biggest companies in the world and they fully agreed it had to be taken in confident steps. It's the way anybody would have developed it."

Sir Arvi says no-one at that time had developed a mine with all the challenges of Olympic Dam. "There was nothing quite like it anywhere else — the mineralogy, the size of it, the circumstances and the political surroundings, the availability of water, the Aboriginal issues, so it was unique in that sense. Just about everywhere else, we could say someone had done something like that previously. For example, when we developed Alcoa of Australia, there were plenty of aluminium companies in the world that had developed large alumina projects, large smelting projects, it wasn't unique. There was a template and it was just a matter of doing it. But there was nothing to follow for developing Olympic Dam."

Sir Arvi says he never doubted that Olympic Dam at some stage would prove to be a really important asset, "but I can't say that I could foresee what happened or how it happened (in prices and markets). We never thought in our wildest dreams that copper would be $3 or $4 a pound. If anyone had suggested that we would have thought they were mentally affected," he says with a laugh.

Sir Arvi says BHP Billiton, unlike Western Mining, can afford to wait for the best time to develop Olympic Dam. "They are big enough, they have cash flow from all sorts of places and more than they need. They have no problem in allocating capital, so they are in that sense an ideal company to own Olympic Dam. A smaller company would not be in the same comfortable situation. I guess the answer is that very large orebodies need very large companies to develop them."

Perhaps Western Mining was too small for Olympic Dam, but it's highly unlikely any other company would have discovered it in the 1970s or had the conviction and persistence to develop it over so many difficult decades.

Western Mining delivered BHP Billiton the world's largest mineral deposit just in time for the world's biggest resources boom. The new owner's expansion plans show it certainly recognises the value of what it inherited, but let's hope BHP Billiton also appreciates what it took to discover Olympic Dam. For Western Mining's greatest legacy was to demonstrate that incredible orebodies exist deep down in the Earth's crust, and that with innovation, persistence and courage they can be discovered and brought into production. As time goes on, the world's supply of minerals will increasingly depend on such deeper deposits. The challenge for present and future generations in the minerals industry will be to understand those lessons and discover many more Olympic Dams.

BHP Billiton's immediate challenge is to finalise plans and obtain approval for its massive expansion of the mine. The company has many factors on its side, including the step-change in the growth of China and India and soaring demand for uranium around the world. BHP Billiton even has the benefit of more benign community attitudes towards uranium and nuclear power. For example, it plans to rail freight much of the additional production as concentrate to Darwin for export to purpose-built processing plants, mostly likely in China. The trains will carry 1.6 million tonnes of copper concentrate every year with uranium still contained. In the 1980s, this would been unthinkable — even if dedicated custom smelters were available to take the material — because of the opportunity created for anti-uranium protests aimed at closing down production.

Even with all this going for it, BHP Billiton's plans still require heroic levels of courage and patience. The expansion is expected to cost anywhere between $20 billion and $40 billion, and will involve up to seven years of digging worthless overburden before new ore reserves can be mined. Olympic Dam is still laying down challenges, even for the world's largest resources company. There will surely be many other challenges to overcome before we reach the final end of the Olympic Dam story in a century or two.

INDEX

A

Aborigines, 55, 170

Acropolis (also Appendicitis Dam), 16, 85, 90, 95, 98, 103-106

Adam, Robin, 135

Adelaidean sediments, 79, 81-83, 85, 91, 106, 108-112, 127

Alcoa of Australia, 36, 40, 123, 150, 170

Alcoa, 41, 48

Allison, T. (Tom), 1, 95, 97, 103, 105, 111, 146, 147

alumina, 35, 48, 123, 150, 170

Anaconda, 137

Andamooka, 16, 76, 77, 79, 80-86, 90, 93, 109, 148

Andamooka Precious Stones Field, 86, 91, 111

Andamooka Station, 147, 148

Andina, 12, 13

Anglo American, 77

Annan, R. (Robert), 32, 33

Appendicitis Dam (also Acropolis), 16, 85, 90, 95, 98, 103-106

Archean, 49, 66

Arcoona, 91, 113, 125

Armstrong, F. (Florence), 42

Australasian Institute of Mining and Metallurgy, 151

Australian Democrats, 155-157

Australian Mineral Development Laboratories (AMDEL), 72, 120, 143

Australian National University, 59

B

Baillieu, W.L. (William), 28, 29

Ballarat, 75, 79, 82, 108, 116, 120

Bannon, J. C. (John), 159, 160

Barr, David, 76

basalt, 3, 21-23, 55, 58-62, 64, 65, 69, 79, 80, 88, 109, 127

Beda Bore, 79

Bendigo, 31, 37

BHP Billiton, 7, 8, 11-15, 21, 25, 28-29, 52, 128, 130, 131, 135-137, 150, 167, 169, 170-171

Billiton, 131

Bill's Lookout, 85, 86, 91

Bopeechee, 91, 113, 125

bornite, 109

Bowen Basin, 13

Bowman, P. (Pearce), 166

BP, 2, 6, 135-141, 145, 148-150, 155, 160, 162, 170

breccia, 18, 21, 110, 149

Brodie-Hall, L. (Sir Laurence), 39, 41, 43

Broken Hill, 9, 27-29, 35, 73, 82

173

Broken Hill Associated Smelters Ltd, 29
Broken Hill South, 29, 54, 107
Brooks, T. (Terry), 95, 96, 98, 112, 119
Bullfinch Mine, 38, 40
Bureau of Mineral Resources, 48, 54, 61, 81, 83, 88, 92
Burke, R. O'Hara, 97
Burmine, 131
Burra, 55

C

Cameron, E. (Eric), 52
Campaign Against Nuclear Energy (CANE), 162
Campbell, D. (Donald), 43, 47, 48, 50, 57
Campbell, I. (Ian), 57
Canada, 33, 38, 46, 47, 49, 56, 66, 67, 114
Carter, D. (Derek), 131
Carrapateena, 94, 131, 132
Case, Pomeroy & Co., 32
Case, W. (Walter), 32, 33
Central Mining and Investment Corporation, 32
chalcocite, 108, 109
chalcopyrite, 108, 109
Chamber's Creek, 97
Chappell, B. (Bruce), 59
Chile, 12, 13, 17, 167
China, 7, 140, 169, 170
Chernobyl, 168
Churchill, W. (Wally), 86

Churchill, W. (Winston), 30
Chuquicamata, 12, 13
Clark, G.C.L. (Sir Lindesay), 10, 27, 34, 39, 41, 48, 123
coal, 13, 35, 126
Codelco, 12
Collins House group, 29, 31
Connolly, T. (Terence), 33
Connor, R.F.X. (Rex), 6
Conzinc Rio Tinto (also CRA), 25, 29
Coolgardie, 57
Cooper Basin, 5, 94, 137
Copping, J. (John), 166
Corcoran, D. (Des), 143
Coronation Dam, 95, 98, 99
Cox's Find, 33
CRA (also Conzinc Riotinto Australasia), 25, 29
craton, 18, 22-24
Crew, R. (Robert), 166
Cripple Creek Mine, 20
Cross, K. (Ken), 20
Cultana, 84

D

Darling Range, 36, 48, 123
Depot Creek, 82, 83
Darwin, 97, 171
de Bavay, A. (Auguste), 28
Dickinson, B. (Sir Ben), 6, 92, 136, 143
digenite, 109
Dixon, G. (Grahme), 140

Downhole logging, 119
Draper, N. (Norm), 105
Dunbar, G. (Gordon), 57
Duncan, D. McP. (David), 81, 109, 129
Duncan, I. (Ian), 166
Dunstan, D. (Don), 5, 136, 137, 143, 144
Durham University, 69

E
Ehrig, K. (Kathy), 19, 130
El Teniente, 12, 13
Emerson, J. (John), 4, 105, 106, 113, 115, 122, 145
Espie, F. (Frank) snr, 39, 47
Estonia, 10, 39, 135
Evans, D.F. (Dan), 51, 66-68, 76, 79, 81-83, 85, 87-91, 93, 95, 99, 100, 102, 105-110, 114, 120, 127, 145
Everett, R. (Rod), 148
Exxon, 137
Eyre Peninsula, 78, 82

F
Flagstaff Hill, 62, 72, 82, 106, 108, 109, 120, 145
Flinders Ranges, 22, 77, 79, 80, 82
Fortescue Copper Project, 56
Foster, N. (Norm), 5, 157-159
Francis, B. (Brian), 152
Fraser, M. (Malcolm), 5, 144-145, 159

G
Gawler Craton, 18, 19, 21-23, 55, 79-81, 93, 94, 106, 109, 128, 149
Gawler Range Volcanics, 21
geophysics, 51, 54, 61, 68-71, 81, 90-94, 128
Geoscience Australia, 61, 64, 130
Geraldton, 76
Germany, 10, 38, 160
Gold Exploration and Finance Company of Australia Ltd, 31
Gold Mines of Australia Ltd, 27, 31, 33, 37
Goldsworthy, R. (Roger), 157
Gomez, R.M. (Rudy), 132
Gottliebsen, R. (Robert), 25, 145
granite, 3, 18, 21, 23, 69, 110, 112, 128
Grasberg Mine, 20
Great Depression, 10, 29, 31, 45, 46
Greenfield, D. (Dave), 101
Guerin, B. (Bruce), 143
Gustafson, J. (John), 33

H
Hamersley Range, 56, 58, 61, 65, 68
Hannington, T. (Ted), 139
Harvard University, 33, 50
Hawke, R.J.L. (Bob), 52, 159, 160
Haynes, D.W. (Douglas), 3, 16, 17, 20, 21, 23, 54, 56-69, 76-85, 88, 100, 107-109, 112, 114-116, 122, 125, 128-130
hematite, 55, 58, 68, 108, 109, 112, 122, 128

Hiltabta Suite, 22
Hollinger Mine, 33
Homestake Mine, 33
Hoover, H. (Herbert), 29
Hudson, G. (Geoff), 57
Hudson, H. (Hugh), 111

I

igneous, 17, 18, 21-23, 57, 63
Imperial Smelting, 32
Indenture Agreement, 5, 155, 157-159
Induced polarisation, 54
IOCG, 17
iron ore, 20, 37, 53, 58, 76, 92, 122, 126, 134

J

Japan, 37, 47, 53, 137, 160
Johns, R.K. (Keith), 92, 94, 111, 143
Johnston, P. (Peter), 166
Joplin, G. (Germaine), 63
JORC Code, 12

K

K/2792 report, 76, 78, 84, 90, 99
Kalgoorlie, 27, 29, 31, 32, 33, 39, 40-42, 47, 50, 52, 57, 67, 70, 73, 76, 88, 101, 112, 119, 125, 147, 153, 155, 165
Kambalda, 25, 36, 41, 49, 56, 57, 70, 102, 112, 115, 123, 134, 147
Kanmantoo Trough, 82
Kangaroo Island, 82
Keweenaw Peninsula, Michigan, 55

Kimberley region, 48, 53, 63
Kinhill, 148
King, H. (Haddon), 33
Kingoonya, 84
Kloppers, M. (Marius), 7, 11, 12
komatiite, 49
Koolanooka, 37, 76, 134
Korean War, 35
Krakatoa, 20
Kwinana, 42, 155

L

Lake Torrens, 83, 85, 132
Lake Eyre, 79
Lalor, J.H. (Jim), v, 1, 3, 10, 51, 60, 66, 75-77, 83, 85, 86, 90, 98, 108-110, 115, 120, 122, 124, 129, 135, 145
lava, 18, 20, 21, 49, 61
Laverton, 147
Lawrence, I. (Ian), 166
Laws, V. (Victor), 32
Leonora, 50
Lissiman, J. (Jim), 105
Loton, B. (Brian), 135

M

maar volcano, 20
magma, 18, 20
magnetite, 54, 58, 68, 69, 128
malachite, 77, 98, 129
mantle plumes, 22, 23
Maree, 82
Marino, 38

Massey-Greene, W. (Sir Walter), 34
McEwen, J. (Sir John), 35
McGill University, 66, 89
McKenzie, B. (Brian), 89
McKinsey & Co., 25
McKinstry, H. (Hugh), 33
McLaughlin, D. H. (Donald), 33
McPhar Geophysics, 54
Meekatharra, 50, 67
Metals Exploration, 131
Milne, L. (Lance), 155, 157
Milton, B. (Bernie), 92, 95, 100
Minorco, 140
Minotaur Resources, 131
Mobil, 137
Moonta, 54, 57, 61, 63, 66, 92, 101
Morgan, 82, 142
Morgan, H. (Hugh), 134-139, 157, 160, 161, 163, 166
Morgan, J. (John), 39
Morgan, W. M. (Bill), 37, 40, 42, 49
Morley, D. (Don), 135, 137
Mt Gunson, 80-86, 90
Mt Coolon, 31, 37
Mt Isa, 7, 11, 62, 93, 129, 154
Mt Lofty Ranges, 79, 80, 82
Muller, H. (Henry), 161, 166
Murray River, 82, 147, 149
Myall Creek, 115

N

natural gamma radiation, 119
NASA, 59
Nelson, R. (Reg), 93
Nevoria Mine, 40
New Broken Hill, 39
New Consolidated Goldfields Ltd, 31, 32
nickel, 4, 8, 11, 25, 36, 41, 42, 47, 49, 50, 53, 56, 58, 67, 70, 110, 112, 119, 123, 155
Norilsk, 8, 11, 12, 13, 14, 16
North Broken Hill, 29, 54
North West Shelf, 126
Northern Australian Craton, 18
Northern Earthmovers, 148
Nullarbor Plain, 101

O

O'Conner, D. (David), 114
O'Driscoll, E.S.T. (Tim), 72, 73, 75, 81, 90, 109, 129, 130, 132
oil, 3, 6, 35, 67, 92, 115, 117, 123, 132, 135, 150
Olary, 82
Oliver, J. (John), 166
Oodnadatta, 4, 102
Outokumpu, 161
OZ Minerals, 131

P

PACE, 94, 132
Palmer, A. (Tony), 166
Paraburdoo, 107

Parbo, A.H. (Sir Arvi), 1, 3, 7, 9, 16, 32, 35, 37-44, 115-117, 123, 124, 134, 137, 139, 144, 150, 156, 158, 160, 165, 169

Parry, K. (Keith), 153, 166

Peake-Denison Range, 102

pegmatite, 57

Pennell, M. (Monty), 135

Perry, B. (Bob), 148

Perth Mint, 37

Peters, T. (Trevor), 166

phreatic, 20

phreatomagmatic, 20

Pilbara Craton, 18

PIRSA (Dept of Primary Industries and Resources South Australia), 94, 130, 132

Pitjantjatjara, 55

plate tectonics, 18, 20, 22, 24, 74, 130

Playford, T. (Sir Thomas), 92

point resistivity, 119

Port Augusta, 77, 79, 80, 82, 83, 94, 95, 97, 98, 102, 147

Port Pirie, 77

Portland, 150, 151

Poseidon, 49

potassium, 119, 120

Pre-Cambrian, 47

Preston, 70, 86, 95, 98, 100, 110

Prominent Hill, 10, 131, 132

Proterozoic, 62, 63, 65, 75, 81, 82

Q

Queen's University, 114

R

Rann, Mike, 143

Reeve, J. (Jim), 20, 151, 152

Resource, definition, 12

Reynolds, J. (John), 156-159, 162

Rhodesia, 56

Rickwood, F. (Frank), 135, 137-138

Rio Tinto, 29, 140, 141

Robinson, L. (Lionel), 27, 28

Robinson, W.S., 9, 27, 28-35, 45, 72

Roopena, 23, 79, 81-83, 85, 88, 100

Roxby Downs, 6, 158

Roxby Downs Granite, 18, 21, 22

Roxby Downs Station, 1, 3, 4, 16, 94-96, 98, 101, 103, 114, 124, 146-148

Roxby Management Services, 126, 139, 145

Royal School of Mines, 70

Rutter, H., 69, 70, 81-83, 85, 91-100, 102, 103, 105, 108, 109, 110, 112, 113, 119, 120, 122, 128

S

Sauer, G. (Graeme), 166

Saville, K. (Kym), 135, 139, 140

Schodde, R. (Richard), 13, 14

Selection Trust, 77, 115, 116, 138

self-potential, 119

Severne, B., 77

Shell, 137

Sheirlaw, N. (Norm), 126, 133, 136

Showers, J. (John), 147, 148

silver, 5, 11, 13, 21, 27, 28, 55, 134, 152, 154
Smith, B. (Barbara), 47
Smith, I. (Ian), 166
Smith, J. (Jeff), 139
Soots, S. (Saima), 38, 39
South Australian Department of Mines (& Energy), 6, 8, 63, 65, 68, 79, 82, 85, 91, 92, 93, 94, 95, 106, 111, 119, 129, 136, 143, 147
South Australian Oil & Gas Corporation, 136
Steart, E. (Eric), 4, 115
stratiform copper deposit, 56, 62, 80, 82, 115, 127
Stuart Creek, 97, 111
Stuart, J. McD. (John), 97
Stuart Shelf, 79, 80, 82-84, 87, 89, 91, 93, 95, 113, 115, 119, 125, 127, 128, 130, 131, 132, 133, 138, 145

T

Talnakh, 12, 13
Tarraji River, 48, 53, 54, 56, 68
Teck Cominco, 132
Tectonic lineaments, 72-75, 129, 130
Telecom, 148
Telfer, 102
Tent Hill, 82, 86
Texaco, 137
Thatcher, G. (George), 105
thorium, 119, 120
Three Mile Island, 5, 150, 168

Time Domain Electromagnetics, 70
Tonkin, D. (David), 5, 144, 145, 155, 157, 159
Torrens Hinge Zone, 78, 80, 82, 90
Triglavcanin, A. (Anton), 54, 70
Triton Mine, 33
Truro, 82

U

ultramafic, 18, 21
Union Corporation, 32
University of Adelaide, 38, 39, 93
University of California, Berkeley, 45, 46, 53
University of Manitoba, 66
University of Western Australia, 45, 46, 48, 56, 57, 75
Uro Bluff, 82, 84
Utah International, 137
uranium, 5, 11-13, 14, 15, 16, 17, 21, 52, 55, 58, 67, 117-120, 121, 122, 125, 133, 134-138, 143, 144, 150, 151, 154, 156-162, 169

V

Viner, I. (Ian), 144

W

Wallaroo, 54, 55, 66
Warburton, 55, 56, 58, 59, 61, 62, 64, 65
Webb, B. (Bruce), 85, 93, 94, 111
Whenan, T. (Ted), 101, 102, 103, 105, 106, 107, 110
Whenan, S. (Shirley), 101, 102

Whenan Shaft, 14, 153, 161

White, A. (Allan), 59

White, G. (George), 147, 148

Whitlam, G. (Gough), 5, 6, 136

Whyalla, 79, 82, 84, 87, 92, 115

Wilmshurst, R. (Ron), 143

Wiluna, 50

Windarra, 49

Witwatersrand, 12

Wirrda Well, 16

Wills, W.J., 97

Wise, C. (Colin), 135

Wittenoom, 50, 56

Woodall, R. (Roy), 1, 3, 10, 17, 43, 45-52, 53-56, 59, 60, 64, 66, 68, 70, 72, 74, 90, 103, 110, 113, 115, 129, 145, 153

Wooltana, 77, 82

Woomera, 83, 84, 103, 111, 113, 149

Woomera Prohibited/Restricted Area, 96, 147

World War I, 10, 29, 32

World War II, 38, 47, 53

X

Xstrata, 7

Y

Yeelirrie, 52, 117, 135, 156

Yilgarn, 38

Yilgarn Craton, 18, 47, 66

Z

Zinc Corporation, 28, 29, 31, 35, 39